Water Management, Food Security and Sustainable Agriculture in Developing Economies

This book addresses strategies for food security and sustainable agriculture in developing economies. The book focuses primarily on India, a fast-developing economy whose natural resource base is not only under enormous stress, but also complex and not amenable to a uniform strategy. It critically reviews issues which continue to dominate the debate on water management for agricultural and food production.

The book examines, using global and national datasets, the validity of the claim that large water resource projects cause serious social and environmental damage. It then explores the potential of these systems for tackling groundwater mining, sustaining well irrigation and reducing the energy footprint of irrigated agriculture through return flow recharge in the command areas. The authors examine claims that the future of Indian agriculture is in rain-fed farming supported by small water harvesting. They question whether water-abundant eastern India could, through a groundwater revolution with the right policy input, become the granary of India. In the process, they look at the less researched aspect of the food security challenge, which is land scarcity in eastern India.

The book analyzes the physical, economic and social impacts of large-scale adoption of micro-irrigation systems, using a farming system approach for north Gujarat. Through an economic valuation of the multiple benefits of tank systems in western Odisha, it shows how value of water from large public irrigation systems could be enhanced. The book also looks at the reasons why the much-needed institutional reforms in canal irrigation have had only limited success in securing higher productivity and equity, using the case studies of Gujarat, Madhya Pradesh and Maharashtra. Finally, it addresses how other countries in the developing world, particularly sub-Saharan Africa, could learn from the Indian experience.

M. Dinesh Kumar is Executive Director of the Institute for Resource Analysis and Policy (IRAP), Hyderabad, India.

M. V. K. Sivamohan is Principal Consultant for IRAP, and is based in Hyderabad.

Nitin Bassi is a Senior Researcher with IRAP, and is based in Delhi.

Water Management, Food Security and Sustainable Agriculture in Developing Economies

Edited by
M. Dinesh Kumar,
M. V. K. Sivamohan
and Nitin Bassi

First published 2013
by Routledge
2 Park Square, Milton Park, Abingdon, Oxfordshire OX14 4RN

Simultaneously published in the USA and Canada
by Routledge
711 Third Avenue, New York, NY 10017

First issued in paperback 2014

Routledge is an imprint of the Taylor & Francis Group, an informa business

British Library Cataloguing in Publication Data
A catalogue record for this book is available from the British Library

Library of Congress Cataloging-in-Publication Data
Water management, food security and sustainable agriculture in developing countries / edited by M. Dinesh Kumar, M.V.K. Sivamohan and Nitin Bassi.
p. cm.
"Simultaneously published in the USA and Canada"–T.p. verso.
Includes bibliographical references and index.
1. Water-supply–India–Management. 2. Water resources development–India. 3. Food security–India. 4. Sustainable agriculture–India. I. Dinesh Kumar, M. II. Sivamohan, M. V. K. III. Bassi, Nitin.
TD303.A1W347 2012
338.10954–dc23
2012014267

ISBN 978–0–415–62407–7 (hbk)
ISBN 978–1–138–90051–6 (pbk)
ISBN 978–0–203–10487–3 (ebk)

Typeset in Baskerville
by Swales & Willis Ltd, Exeter, Devon

We dedicate this volume to the millions of undernourished children in the developing world.

Contents

List of figures ix
List of tables xi
List of contributors xiv
Acknowledgements xvi
List of acronyms and abbreviations xvii

1 **Food security and sustainable agriculture in developing economies: the water challenge** 1
M. DINESH KUMAR, M. V. K SIVAMOHAN AND NITIN BASSI

2 **Key issues in Indian irrigation** 17
M. DINESH KUMAR, A. NARAYANAMOORTHY AND M. V. K. SIVAMOHAN

3 **Food security challenges in India: exploring the nexus between water, land and agricultural production** 38
M. DINESH KUMAR, M. V. K SIVAMOHAN AND A. NARAYANAMOORTHY

4 **Redefining the objectives and criteria for the evaluation of large storages in developing economies** 61
ZANKHANA SHAH

5 **Sector reforms and efficiency improvement: pointers for canal irrigation management** 76
NITIN BASSI, M. DINESH KUMAR AND M. V. K. SIVAMOHAN

6 **Rebuilding traditional water harnessing systems for livelihood enhancement in arid western Rajasthan** 97
NITIN BASSI AND V. NIRANJAN

7 The hydrological and farming system impacts of agricultural water management interventions in north Gujarat, India 116
O. P. SINGH AND M. DINESH KUMAR

8 Technology choices and institutions for improving the economic and livelihood benefits from multiple-use tanks in western Odisha 138
M. DINESH KUMAR, RANJAN PANDA, V. NIRANJAN AND NITIN BASSI

9 Future strategies for agricultural growth in India 164
M. DINESH KUMAR, V. NIRANJAN, A. NARAYANAMOORTHY AND NITIN BASSI

10 Investment strategies and technology options for sustainable agricultural development in Asia: challenges in the emerging context 184
P. K. VISWANATHAN, M. DINESH KUMAR AND M. V. K. SIVAMOHAN

11 Water management for food security and sustainable agriculture: strategic lessons for developing economies 208
M. DINESH KUMAR

Index 229

Figures

2.1 Per capita groundwater withdrawal rates in different states (m^3/annum) 20
2.2 Wellgroundwater irrigated area nexus 21
2.3 Degree of dependence of cities on reservoirs 30
2.4 Net surface and well irrigated area in India (1950–51 to 2006–07) 31
2.5 Intensity of groundwater use in different states 33
3.1 Production of food grains vis-à-vis irrigation in Indian states (2007–08) 41
3.2 Per capita cropping in different states 42
3.3 Per capita groundwater withdrawal rates in different states (m^3/annum) 50
4.1 Dam height vs. storage volume 64
4.2 Dam height vs. reservoir area 65
4.3 Submergence area vs. population displaced 66
5.1 Irrigation potential created and utilized under major and medium irrigation schemes in India, by plan 78
5.2 Equity in canal water distribution, Dharoi irrigation project, Gujarat 82
5.3 Irrigation revenue recovery by year, Madhya Pradesh 85
5.4 Relationship between gross irrigated area and water charges recovery, Satak tank project 87
5.5 Irrigation water charges recovery and O&M cost, Maharashtra 89
5.6 Water storage at Waghad dam reservoir and net irrigated area at Ozar village, Maharashtra 90
5.7 Financial performance of Ozar WUA, Nashik district, Maharashtra 91
6.1 Annual rainfall and moving average (1957–2009), Jaitaran, Pali 98
6.2 Relation between annual rainfall and rainy days 99
6.3 The rise in water level for observation wells under the influence of the khadin near Balada village 104
6.4 Capacity increase of village ponds (by volume), ACF's Rabariyawas and Mundwa locations 106
6.5 The rise in water level for observation wells recharged through Balada pond 107
6.6 Change in water source for households located near Balada pond 108
8.1 Monthly rainfall, potential evaporation and runoff (mm) in Mahanadi basin, upstream of Hirakud 141

8.2 Conceptual model for analyzing surplus value product from the direct use of tank water under various scenarios of water allocation 143
9.1 Net area irrigated by open wells, tube wells and total well irrigated area, AP 172
10.1 Trends in the agricultural exports of major Asian countries 188
10.2 Trends in per capita food production in major countries, 1994–2008 191
10.3 Long-term trends in the per capita net availability of food grains in India 192
10.4 Trends in public and private sector gross capital formation in agriculture in India, 1990–91 to 2008–09 195
10.5 Trends in farm mechanization in India and other countries, 1975–2007 196
11.1 Global Hunger Index vs. SWUI 214

Tables

2.1 Percentage of dug wells and deep tube wells suffering from poor discharge in selected Indian states 19
2.2 Gross irrigated area and well irrigated area for major Indian states 20
2.3 Reduction in average command area of wells over time in selected districts of Madhya Pradesh 22
2.4 Per capita renewable water resources and per capita water demand in agriculture in two river basins 23
2.5 Average reference evapo-transpiration against annual water resources in selected river basins in water-scarce regions 24
2.6 Estimated unit cost of artificial recharge structures built under CGWB pilot scheme 27
3.1 Well failures in different categories from eight major Indian states (2001) 46
3.2 Percentage of dug wells and deep tube wells suffering from poor discharge in selected Indian states 47
3.3 Reduction in the average command area of wells over time in the Narmada basin, Madhya Pradesh 47
3.4 Gross irrigated area and well irrigated area for major Indian states 49
3.5 Aggregate and relative contribution of states falling in semi-arid alluvial areas to India' wheat and rice production (2000) 51
3.6 Contribution of states falling in hard rock areas to India's (rough) rice production in 2006 52
4.1 Large dams in India 63
5.1 Seasonal irrigation water rates for some crops, Gujarat 81
6.1 Population, land use and groundwater availability in western Rajasthan 101
6.2 Crop economics, inside and outside the water spread area of the khadin 104
6.3 Benefit–cost ratio for the Baladakhadin 105
6.4 Water use pattern for households located near Lakholav pond 109
6.5 Expenditure pattern pre- and post-pond rehabilitation 109
6.6 Seasonal increased time available, within the benefited households, to undertake various works 111

7.1 Per capita renewable freshwater availability in Gujarat, by region 118
7.2 Key physical achievements of the North Gujarat Initiative 119
7.3 Average family size of adopters and non-adopters of water-saving technology 122
7.4 Average farm holdings of adopters and non-adopters of water-saving technology 122
7.5 Irrigation water use for different crops before and after adoption of MI 123
7.6 Yield of irrigated crops with and without MI systems 125
7.7 The area under different crops of adopters before and after adoption of MI 126
7.8 Yield and gross income obtained by farmers from different types of livestock before and after adoption of MI systems 127
7.9 The net income, modified net income and water productivity, in physical and economic terms, with and without the adoption of MI systems 128
7.10 Benefit–cost analysis of MI systems for different crops 130
7.11 The impact of adoption on farm income 131
8.1 The area under different crops irrigated from tank water in five selected villages during a normal year 148
8.2 The area under different crops irrigated from tank water in five selected villages during a drought year 149
8.3 Yield of crops irrigated by tank water in the five selected villages during a normal year 149
8.4 Yield of crops irrigated by tank water in the five selected villages during a drought year 150
8.5 Net returns from crops irrigated by tank water in the five selected villages during a normal year 150
8.6 Net returns from crops irrigated by tank water and, also, upland in the five selected villages during a drought year 151
8.7 The total economic value of wetland irrigation from the five selected tanks 152
8.8 Quantum of catch of different varieties of fish from the five selected tanks, and the market value of the catch 154
8.9 The extent of the use of tank water for domestic and recreational needs 154
8.10 Estimated public cost of the water supply in the villages in Sambalpur: two scenarios 156
8.11 Value of the social good and recreational service provided by the tanks 156
8.12 The economic value of various uses of water during normal and drought years 157
8.13 The water productivity of different crops during normal and drought years 158
10.1 Trends in the production of food grains, 1961 to 2009 187

10.2 Trends in agricultural GDP and its sectoral composition across major countries, 1985 to 2010 189
10.3 Trends in per capita food production in major countries 191
10.4 Annual freshwater withdrawal for major uses, 2009 193
10.5 Trends in agricultural subsidies in India, 2000–01 to 2008–09 196
10.6 Food production increments required in Asia by 2015 (per cent) 199
10.7 The key areas for investment, the strategic interventions/policies and the key challenges 202

Contributors

Nitin Bassi is a Senior Researcher at the Institute for Resource Analysis and Policy, Delhi office and previously worked with the International Water Management Institute. He holds an M. Phil in natural resource management from IIFM and has over six years' experience undertaking research in the field of water resource management and farmer participation. He has published many papers in national and international journals.

M. Dinesh Kumar is Executive Director at the Institute for Resource Analysis and Policy, Hyderabad. Dr. Kumar holds a PhD in water management, an ME in Water Resources Management and a BTech in civil engineering. Dr. Kumar is a senior water resources scientist with over two decades of experience working on technical, economic, institutional and policy related issues in water management in India. He has published in several international and national journals on water and energy, and has three books, one edited volume and over 125 research papers to his credit. Currently, he is also Associate Editor of *Water Policy*, an international peer-reviewed journal published by IWA.

A. Narayanamoorthy is NABARD Chair Professor and Head of the Department of Economics and Rural Development, Alagappa University, Karaikudi, Tamil Nadu, India.

V. Niranjan is a Research Officer at the Institute for Resource Analysis and Policy, Hyderabad, and holds an MTech in Environment Management from Jawaharlal Nehru Technological University, Hyderabad.

Ranjan Panda is convener of the network Water Initiatives Odisha, which is at the forefront of action and advocacy on water, environment and climate change in Odisha.

Zankhana Shah holds an MPhil in Environment Policy from the University of Cambridge, an MTech in Urban and Regional Planning from CEPT University and a Master's Degree in Social Work from Sardar Patel University. She is presently working as a Junior Carbon Consultant in Energize, Cambridge, UK.

O. P. Singh is Assistant Professor, Department of Agricultural Economics, Institute of Agricultural Sciences, Banaras Hindu University, Varanasi.

M. V. K. Sivamohan is Principal Consultant to the Institute for Resource Analysis and Policy, Hyderabad. He has over 40 years of research, consultancy and training experience. Formerly a Senior Member of the Faculty and Area Chairman (Agriculture and Rural Development) at the Administrative Staff College of India, Hyderabad, Dr Sivamohan holds a PhD in the social sciences. He has worked with several international organizations, such as the Irrigation Research Group, Cornell University (USA); the Natural Resource Institute (UK); and the International Water Management Institute (India). He has several publications in both national and international journals and has four edited volumes to his credit. His specialization is in natural resources management and public administration.

P. K. Viswanathan is Associate Professor at the Gujarat Institute of Development Research, Ahmedabad. Viswanathan has a PhD. in Economics from the Institute for Social and Economic Change, Bangalore.

Acknowledgments

This volume deals with issues common to a number of Asian and African countries: water, sustainable agriculture and food security issues. The editors are grateful to all of the contributors to this volume, for asserting fresh arguments and ideas about one of the most complex challenges confronting the developing countries of the world today.

Thanks are also due to the Sir Ratan Tata Trust (SRTT), Mumbai, and the IWMI-Tata Water Policy Research Program, Hyderabad, which have supported the Institute for Resource Analysis and Policy, Hyderabad, in undertaking the studies from which two chapters of this book (Chapters 7 and 8) are taken. The editors are also grateful to three anonymous reviewers who have offered very constructive and valuable comments on the scope of the book and some of the chapters in the draft version of the manuscript. The compilation and editing of this academic work required significant co-ordination and orchestration on the part of the editors and contributors, and the comments received helped immensely in sharpening the focus of the chapters.

The editors are indeed grateful to Mr Tim Hardwick, Senior Commissioning Editor, and Ms Ashley Irons, Editorial Assistant, both of Earthscan, as well as to the team at Routledge (part of the Taylor and Francis Group), for their prompt and helpful suggestions in bringing out this volume.

Acronyms and abbreviations

ACF	Ambuja Cement Foundation
ADB	Asian Development Bank
AKRSP-I	Aga Khan Rural Support Programme (India)
AP	Andhra Pradesh
BC Ratio	Benefit–Cost Ratio
BCM	Billion Cubic Metres
CAD	Command Area Development
CAZRI	Central Arid Zone Research Institute
CBIP	Central Board of Irrigation and Power
CF	Consumptive Fraction
CGWB	Central Ground Water Board
CROPWAT	A computer programme devised by the Food and Agriculture Organization of the United Nations, for the estimation of crop water requirements and irrigation requirements, based on climate and crop data
Crore	1 crore equals 10 million Indian rupees
CWC	Central Water Commission
DFID	Department for International Development
DSC	Development Support Centre
ECOBEN	Economic Benefit
ET	Evapo-transpiration
ET_0	Reference Evapo-transpiration
EU	Emission Uniformity
FAO	Food and Agriculture Organisation of the United Nations
FV	Flow Variation
GBM	Ganga-Brahmaputra-Meghna
GCA	Gross Cropped Area
GDP	Gross Domestic Product
GEC	Gujarat Ecological Commission
GHARP	Great Horn of Africa Rainwater Partnership
GHI	Global Hunger Index
GOG	Government of Gujarat
GOI	Government of India

GOM	Government of Maharashtra
GOO	Government of Odisha
GR	Government Resolution
GSDA	Groundwater Survey and Development Agency
Ha/ ha	Hectare
HDR	Human Development Report
HHs	Households
HP	Horsepower
ICAR	Indian Council of Agricultural Research
ICEF	Indo-Canada Environment Facility
ICOLD	International Commission on Large Dams
IDE	International Development Enterprises
IFPRI	International Food Policy Research Institute
IIPA	Indian Institute of Public Administration
IMPACT	International Model for Policy Analysis of Agricultural Commodities and Trade
IMT	Irrigation Management Transfer
INR	Indian Rupee
IPC	Irrigation Potential Created
IPU	Irrigation Potential Utilized
IRAP	Institute for Resource Analysis and Policy
IRMA	Institute of Rural Management Anand
IWMI	International Water Management Institute
KARI	Kenya Agricultural Research Institute
Kg/kg	Kilogramme
km^3	Cubic Kilometre
kWh	Kilowatt hour
Lac	1 lac equals 1,00,000 (100,000 in the Western numbering system); 1 lac hectare equals 1 billion square metres
Lpcd	Litre per capita per day
Lt/lt	Litre
M	Metre
M ha m/M ham	Million Hectare Metre
M & R	Maintenance and Repair
M.tonne	Million Tonnes (metric, unless stated otherwise)
m^3	Cubic Metre
MAF	Million Acre Feet
MCM	Million Cubic Metres
MI	Micro Irrigation
MI	Minor Irrigation
MJ	Million Joules
Mm	Millimetre
MOU	Memorandum of Understanding
MP	Madhya Pradesh
MSSRF	MS Swaminathan Research Foundation

MUS	Multiple Use Systems
NA	Not Applicable
NCAER	National Council of Applied Economic Research
NCIWRD	National Commission on Integrated Water Resources Development
NGI	North Gujarat Initiative
NGO	Non-Governmental Organization
NI	Net Income
NIA	Net Irrigated Area
NREGS	National Rural Employment Guarantee Scheme
NRAA	National Rain-fed Area Authority
PET	Potential Evapo-transpiration
PIF	Public Interest Foundation
PIM	Participatory Irrigation Management
PPP	Public–Private Partnership
Qt/ qt	Quintal
RWHS	Rain Water Harvesting System
RWH	Rain Water Harvesting
SC	Scheduled Caste
SMS	Short Messaging System
SOFILWM	Society for Integrated Land and Water Management
SOPPECOM	Society for Promoting Participative Ecosystem Management
Sq. km/km^2	Square kilometre
Sq. m/m^2	Square Metre
SSP	Sardar Sarovar Project
ST	Scheduled Tribe
SWUI	Sustainable Water Use Index
TEV	Total Economic Value
TFP	Total Factor Productivity
TMC	Thousand Million Cubic Feet
TMI	Traditional Method of Irrigation
TOI	The Times of India
TP	Treadle Pump
UNDP	United Nations Development Programme
UNICEF	United Nations Children Fund
UP	Uttar Pradesh
US	United States
USA	United States of America
WALMI	Water and Land Management Institute
WFP	World Food Programme
WH	Water Harvesting
WST	Water Saving Technology
WUAs	Water Users' Associations

1 Food security and sustainable agriculture in developing economies

The water challenge

M. Dinesh Kumar, M. V. K. Sivamohan and Nitin Bassi

Introduction

Sustainable growth in agriculture is crucial for food security, rural development and the long-term economic growth of many poor nations in Asia and Africa, and it is well known that only those strategies which are built around agricultural growth can take hundreds of millions living in Asia and Africa out of the poverty trap (Schultz, 1979; DFID, 2004; Cervantes-Godoy and Dewbre, 2010). Recent analyses show that countries which face severe problems of hunger are also characterized by lower levels of water security (HDR, 2006; Kumar et al., 2008a; Kumar, 2011). Return on investment in water – in the form of infrastructure, institutions and policies – is expected to be high in countries which are low in the economic growth indices (Grey and Sadoff, 2007). A recent paper on the growth tragedy of East Africa mirrors the correlation between long-term reduction in rainfall and the decline in economic growth rates (Barrios et al., 2010). Analysis using global datasets to show the relationship between water security, food security, human development and economic growth is also available (Kumar et al., 2008a; Shah and Kumar, 2008; Kumar, 2011).

India is an illustrious example of the undisputable role water development has in boosting agricultural development, reducing poverty and maintaining economic growth in developing economies. Since independence, India's per capita net national product recorded a compounded growth rate of 1.7 per cent (Datt, 1997). The contribution of agricultural production to this growth has been significant (Kumar, 2003). Irrigation has been pivotal in enhancing grain production, and ensuring food security at the national level, with two-thirds of agricultural production coming from irrigated areas (Kumar, 2003 based on Evenson et al., 1999; GOI, 1999, 2002).

But this growth has not been uniform. A large portion of the growth in agricultural production has come from the northern region, mainly Punjab, Haryana and western UP, which reaped the benefits of the Green Revolution rather quickly (see Table 21, Evenson et al., 1999). They achieved it by enhancing the use of conventional inputs such as irrigation and fertilizers, and by improving the total factor productivity (TFP) through the adoption of new crop technologies.

There has been significant expansion in irrigated agriculture in the region. Similarly, the annual growth in TFP during the period from 1956 to 1987 was 1.40 per cent for the region, against a national average of 1.13 per cent (Evenson et al., 1999).

The growth rate in TFP is lowest for the eastern region comprising Bihar, Odisha and West Bengal (0.75 per cent). Further, it has declined over three decades (1956–87) from 1.5 during 1956–65 to 0.70 during 1977–87 (Evenson et al., 1999). The grain yields remained the lowest in Bihar (GOI, 1999). There are many reasons for the low agricultural productivity in this region. First is the low level of cultivation and use of irrigation. Irrigation, when compared with population size, is poorest in states like Bihar and Odisha as compared to states like Punjab and Haryana (Kumar, 2003). The situation in Bihar is noteworthy because it has one of the lowest per capita cropped areas (at 0.092 ha); this and the relatively poor irrigation are compounded by abysmal yield levels.

Due to meagre farm surpluses, farmers are not able to invest in irrigation sources, expand irrigation, increase cropping intensities and enhance the crop yields. Though the irrigation potential of groundwater is very high, the pace at which development of groundwater resources takes place in the region is extremely slow. For example, the development of groundwater in Bihar, expressed as a ratio of the gross draft and the replenishable groundwater resources, is only 23.3 per cent (GOI, 1999, p. 16). Poor irrigation also influences the level of use of inputs – such as fertilizers, pesticides and hybrid crop varieties – adversely, resulting in low TFP.

In the water-abundant Indo-Gangetic plains, such as eastern UP, Bihar, and West Bengal, socio-economic deprivation significantly hinders both investment in irrigation development and any increase in the use of inputs for maximizing agricultural output (Shah, 2001; Kumar, 2003). The average per capita income in states such as Bihar and UP is far below the national average (GOI, 2002, p. 35), and the resource-poor, small and marginal farmers in the region prefer buying water from well owners, at prohibitive prices, to irrigate crops (Kumar, 2007), even though it is unviable to do so.

In addition to this, lack of public funds to invest in the water resource development sector is a third challenge to economic growth. The TFP growth in agriculture is relatively high in peninsular India (Evenson et al., 1999), which also has high per capita agricultural GDP. The region is also a major exporter of cereals to agriculturally backward regions that experience food deficits (Amarasinghe et al., 2005). But scarcity of irrigation water is becoming a major impediment to sustaining this growth. Inter-basin transfer of water from water-abundant river basins to those which are water-scarce could help augment the irrigation potential of these regions (Kumar and Singh, 2005; Kumar et al., 2008b), while augmenting the country's water supply potential by 200–250 BCM. However, this may cost as much as 20–25 billion dollars (Chaturvedi, 2000). The availability (or lack thereof) of finance is a major stumbling block for such projects (Kumar, 2003). Issues of political economy, lack of suitable governance mechanisms and rampant corruption add to the challenges.

It is evident that water is important for agricultural growth and rural poverty alleviation. Since independence, there has been a remarkable increase in water supplies for irrigation, through the building of large and medium surface irrigation schemes, and groundwater development. However, most of the major schemes for irrigation had been planned and implemented before more recent major advancements in hydro sciences. As a consequence, the efficiency of utilization of water for irrigation has been extremely low in India, as in many other developing countries (Chaturvedi, 2000). The potential for adoption of water-saving micro-irrigation systems to improve water use efficiency in agriculture appears to be quite insignificant when compared to India's gross irrigated area (Kumar et al., 2008c).

On the other hand, the demand for water in agriculture is growing, both due to the increasing food grain needs of the growing population and the growing preference for water-intensive cash crops. The per capita demand for food grains is growing due to changing consumption patterns. While the average per capita demand for cereals for direct consumption has declined in the recent past, the demand for cereals for animal feed is increasing, owing to increased demand for animal products such as milk, beef, mutton, pork, poultry products, eggs and freshwater fish (Amarasinghe et al., 2007). In the urban and industrial sectors, the growth is more rapid, owing to the faster growth of urban populations and rapid industrialization. By one estimate, the total water requirement for human and animal uses, industrial production and irrigated agriculture would be 104.50 million hectare metres in the year 2025. A comparison of water requirements and utilizable future supplies shows that, by the year 2025, the magnitude of scarcity would be 26.20 million hectare metres (Kumar, 2010).[1] But the scarcity is not going to hit all regions uniformly. The naturally water-scarce regions would be hit adversely, as the demands for water from agriculture, industrial and urban sectors are quite high in these regions, while the renewable water resources available within the region are scarce. This is compounded by the demand for water for reducing environmental water stress in the rivers of these regions. In the absence of proper legal and institutional regimes under which water rights can be allocated among the competing uses, rights will be politically contested, leading to conflicts (Kumar, 2010).

The water utilities of large urban centres in India are highly dependent on water imported from distant reservoirs. Many of these large urban areas are in naturally water-scarce regions (Mukherjee et al., 2010). Urban areas being economically and politically powerful (Banik, 1997), it is very likely that they manage the huge additional supplies required from the rural areas. This can have major implications for irrigated agriculture, especially for the economically weaker groups. When supplies decline due to increasing reallocation of water to other sectors, the wealthier and more influential farmers (often taking advantage of their preferential location within the hydraulic system) draw the lion's share of the remaining limited water, continuing to enjoy as much access as when supplies were in plenty, and thereby depriving less privileged farmers of their rights (Kumar, 2003). Thus, the agriculture sector in naturally water-scarce regions would face severe competition for water from other sectors such as industry, urban drinking and the environment.

While a scarcity of water for the irrigation required to meet increasing demand for cereals, oil seeds and fibre and the micro economic needs of the farming communities are the foremost concerns, variations in the demand–supply balance across regions, and competing claims made by urban domestic, manufacturing and environmental sectors, accentuate the problem remarkably (Kumar, 2010).

In many developing countries of Asia and Africa, rural communities face severe shortages of water for their domestic productive needs (Falkenmark and Rockström, 2004; Grey and Sadoff, 2007; HDR, 2006). In most countries of sub-Saharan Africa, this scarcity of water is economic in nature: the current water utilization is far less than the utilizable water resources due to poor institutional capacity within the water sector (Falkenmark and Rockström, 2004). Public investment in irrigation and rural water supply is crucial if we are to take the hundreds of millions of people in these regions out of malnutrition and hunger and socio-economic backwardness, all of which could happen through domestic water security and the stabilization of agricultural production (Kumar et al., 2008d; Shah and Kumar, 2008).

The dominant views in India's agriculture

The grave situation emerging with regard to agricultural growth, rural development, food security and livelihoods has not been appreciated. Attempts to underplay the gravity of the situation curtail healthy debate on the water management solutions needed for the future. The debates in this context relate to the future role of agriculture and, therefore, irrigation in an economy which is in transition; the potential of land-scarce and water-rich regions to produce surplus food for less endowed regions; supplementary irrigation to enhance rain-fed yields, and rainwater harvesting to provide water for supplementary irrigation; the potential of micro irrigation systems in saving water in agriculture; and the feasibility of introducing water-efficient crops in naturally water-scarce regions to improve water productivity in agriculture.

These are some misconceptions prevailing. Surface irrigation is becoming increasingly irrelevant in India's irrigation landscape, and is highly inefficient (IWMI, 2007). Well irrigation is very productive; it is efficient and its growth can continue in water-abundant eastern India to break the impasse in agricultural growth, reduce rural poverty and bring about prosperity in that region (Shah, 2001; Mukherjee, 2003). Rain-fed farming is very efficient, can contribute greatly (with in situ rainwater harvesting and other measures) to India's agricultural growth in dry land areas, and is capable of becoming the alternative to large public investment in irrigation infrastructure (Shah, 2009). Micro-irrigation and the introduction of water-efficient crops save water in water-scarce regions, with no social costs (Suresh Kumar and Palanisami, 2011). Large dams cause several negative social and environmental effects which outweigh the benefits from them – benefits that are often over-played. Returns from large surface irrigation systems are too few (D'Souza, 2002; Dharmadhikary, 2005). This book provides an alternative way of thinking.

Scope of the book

This book discusses strategies for food security and sustainable agriculture in developing economies whose natural resource base supporting food and agricultural production are not only under enormous stress, but also quite complex and not amenable to a single strategy. India has the second largest rural population and the highest number of agricultural producers, of which the majority are small and marginal farmers. It has over one hundred agro-ecological sub-regions. It consists of water-scarce and water-rich regions. There are regions that are agriculturally as prosperous as some of the agriculturally most advanced regions of the world. Among all countries, it has the highest number of people suffering from food insecurity and malnutrition. Therefore, developing economies elsewhere can learn a great deal from observing which strategies India pursues to address its long-term agricultural food production needs.

The book attempts to address some of the misconceptions which continue to dominate the debate on water management for agriculture and food production. First of all, it examines the validity of the argument that large water resource projects cause serious social and environmental damage. It then explores the potential of these systems in tackling groundwater mining, sustaining well irrigation and reducing the energy footprint of irrigated agriculture through return flow recharge in the command areas. The book inquires into the validity of the claims that the future of Indian agriculture is in rain-fed farming supported by small water harvesting. Analysis of the impact of massive investment in small water harvesting schemes, uncontrolled groundwater development and large water projects in India is used to determine strategies for water resource development in sub-Saharan Africa that are appropriate in reducing negative social welfare effects.

The book also revisits the claim by scholars that water-abundant eastern India could – with the right policy input and a groundwater revolution – become the granary of India. In the process, it also looks at the less researched aspect of the food security challenge, which is the land crisis in eastern India. It also looks at the reasons why there has been only limited success in bringing about the much-needed institutional reforms in canal irrigation (required to secure higher productivity), using the case studies of Gujarat, Madhya Pradesh and Maharashtra.

While examining claims about the potential (and potential benefits) of micro-irrigation systems, the book argues that the criteria for evaluating their advantages are simplistic. The book analyzes the physical, economic and social impact of the large-scale adoption of these systems, with particular reference to the north of the Gujarat region. Some important lessons for sub-Saharan Africa, as to the factors that would determine the adoption of micro-irrigation systems, are highlighted.

In undertaking an economic valuation of the multiple use benefits of tank systems located in western Odisha for irrigation, domestic water use, livestock drinking and fisheries, the authors show how the economic value of water from large public irrigation systems, which is traditionally and dominantly used in irrigated crop production, could be enhanced by converting them into multiple use systems. It also suggests how the concept of multiple use water systems could be used in sub-Saharan Africa to maximize the benefits to be gained from the much-needed future public investment in irrigation and drinking water supply infrastructure.

The analyses show how the current policies and programmes of the government in water and agriculture sector are degenerative, driven by political considerations which promote inequity in access and the inefficient use of what is a precious resource. It argues that those responsible for programmes and policies for agricultural growth in India should embark on a multi-pronged, region-specific strategy, which is rooted in the hydrological, agro-climatic and socio-economic realities of the regions concerned, with the right combination of technical interventions and economic instruments.

Contents of the book

Chapter 2 examines the scientific validity of the claims that canal irrigation is declining; that the investment in surface irrigation does not lead to a proportional increase in size of area irrigated; that well irrigation will take over; and that, in the longer term, surface irrigation will become non-existent in India; that the sustainability of well irrigation in over-exploited regions can be improved through local water harvesting and recharge; and that well irrigation can continue to grow in water-abundant regions. While challenging the very basis for predicting a bleak future for surface irrigation in India, the book provides a counter-argument that surface water resource systems enjoy greater advantages over groundwater irrigation not only from the point of view of the physical transferability of water, but also from the point of view of the capacities required for governance and management of the resource. While contesting the claim that well irrigation is private sector based with little or no government support, the authors go on to argue that subsidized electricity for groundwater pumping produces negative welfare effects. It asserts that subsidies in surface irrigation are justified in many situations because of the positive externalities that such irrigation generates; that current evaluations of the performance of surface irrigation systems are based on the area irrigated, an approach that the authors argue is severely lacking; and, further, that criticism of the poor efficiency of surface irrigation systems are based on obsolete irrigation management concepts which treat the amount of water supplied in excess of the crop water requirement as loss or "wastage".

With the rising demand for food and declining per capita availability, India is facing a food crisis. As the total cereal demand in India is projected to be 291 million tonnes by 2025 (Amarasinghe et al., 2007), this food crisis is only likely to intensify in the future. There are three reasons for this. First, the maximum annual cereal production achieved so far in India has hardly touched 231 million tonnes (in 2007–08). Second, the net area under cultivation and area under food grains has more or less stagnated (GOI, 2008). Third, the cereal yields have not shown any significant growth over time. Expanding the cereal-producing area to enhance production would require additional water for irrigation to increase cropping intensities (Kumar, 2003). On the other hand, meeting the rising demand for milk and other dairy products requires intensive dairy production, which is highly water intensive – especially if undertaken in those semi-arid and arid regions which are currently the major contributors to the country's dairy production. This is because such production would require more irrigated fodder crop (Singh, 2004; Singh and Kumar, 2009) along with commercially produced animal feed.

A few scholars have argued that India's agricultural growth impasse can be breached through a boost in well irrigation in eastern India (Shah, 2001; Mukherjee, 2003; Sharma, 2009). Major arguments about the poverty alleviation impact of groundwater are made in the context of eastern India (Shah, 2001; Mukherjee, 2003; IWMI, 2007). As Mukherjee argues in an IWMI policy brief, "in regions of abundant rainfall and good alluvial aquifers, groundwater irrigation can be a powerful catalyst in reducing poverty" (IWMI, 2007). These arguments have assumed prominence by virtue of the fact that some of the cereal-producing regions have started to face the serious problems of groundwater depletion and water scarcity, causing lower growth in productivity in recent years. Such regions include Punjab and Haryana in the north, and Andhra Pradesh and Tamil Nadu in the south. The theory of virtual water trade has also resulted in less emphasis being put on the magnitude of the food security challenge India is poised to face: it was felt that countries like India should start importing cereals, instead of trying to meet all the demand through domestic production.

Chapter 3 revisits this food security debate. It first reviews the Global Hunger Index (GHI) for India and other developing countries of South Asia and Africa which are food insecure. It then examines India's food security challenge from the point of view of food supply, and explores the future strategies for sustaining food production and agricultural growth. A comparative analysis of naturally water-scarce and naturally water-rich river basins is carried out. In the former areas, the annual agricultural demand far exceeds the renewable water resources, owing to high per capita arable land and aridity. In contrast, the latter are physically water-abundant and subject to high rainfall, with the annual renewable water resources far exceeding the annual water demand in agriculture, which is constrained by low per capita arable land. Because of low productivity and limited arable land, water-rich regions experience food deficit. This leads to increased pressure on land-rich and water-scarce regions, which are agriculturally prosperous, to provide surplus food production.

The authors go on to consider the magnitude of food insecurity caused by groundwater depletion in the agriculturally prosperous, naturally water-scarce, regions, which also export cereals to water-rich regions. The negative consequences of groundwater over-exploitation is assessed, such as well failures, reduction in the yield of wells, the reduction of well commands and the cost of groundwater abstraction, before the impact of depletion on agricultural production and food security is considered. The chapter authors suggest strategies for enhancing agricultural production and ensuring food security, so as to overcome the problems caused by lack of arable land in regions of water abundance, and water scarcity in agriculturally prosperous regions with sufficient arable land. The strategy includes surface water export from land-scarce and water-rich regions to land-rich and water-scarce regions, and improving the productivity of use of water in agriculture in the water-scarce regions.

Large irrigation projects are the target of growing criticism worldwide for the negative social and environmental effect they are likely to cause. Critics argue that the costs outweigh their intended benefits. Large dam projects in particular increasingly face opposition, while their role in development has been largely ignored. In Chapter 4, we revisit the large dam versus small dam debate. India has 4,635 "large dams" as per the ICOLD definition, which considers only technical criteria such as height and storage volume in classifying dams as large. By some estimates, there are 47,000 large dams in the world.

Two issues are considered in Chapter 4. First, what should be the best criterion for classifying large water storages in a way that truly reflects the engineering, social and environmental challenges posed by them? Second, what new objectives and criteria need to be incorporated in the cost-benefit analysis of dams so as to make it comprehensive? The analysis of data on 13,631 large dams from around the world shows that the height of the dam does not have any bearing on the volume of water stored, a strong indicator of the safety hazard posed by dams. Further analysis using data on 9,878 large dams shows that the height has no bearing on the area of land submerged, again an indicator of the negative social and environmental effects. Data on 156 large dams across India shows that a normative relationship exists between the area of submergence and the number of people displaced by a dam. Based on these findings, the authors argue that a combination of criteria such as height, storage volume and the area under submergence should be considered in any assessment of the negative social and environmental consequences of dams. Further analysis shows that the available estimates of dam displacement could be "gross over-estimates": in the order of magnitude of eight.

There are inherent limitations in the benefit–cost analysis of large multi-purpose reservoir projects, which ignore the several positive externalities they produce. The authors illustrate their significant positive impact on stabilizing national food prices, contributing clean energy, improving recharge to groundwater in semi-arid and arid regions, and ensuring social security. They argue that any economic evaluation of such reservoir projects should consider these positive externalities, as well as the direct and economic benefits. The authors estimate that the reduction in food prices resulting from the use of large dams in India is worth around INR 42.90 billion annually. At the same time, the negative externality effects of large dams should be built into the cost of dam projects in order to increase the accountability of water development agencies to those communities adversely affected by large dams.

The Sardar Sarovar Project (SSP) on Narmada River has not been an exception to the growing criticism of large reservoir projects. The limited analyses of the performance of SSP undertaken so far are based on narrow objectives with ideological overtones. The underlying concern is that, as several decades had passed since the project was initially conceptualized and planned, the social, economic and environmental context has undergone a metamorphosis. While water-logging was projected as a negative externality of the project, seepage and return flows from irrigation are found to be contributing to sustaining well yields and reducing the energy requirements for lifting water, thereby inducing a positive externality on society. With increased recharge of water from rivers and canals and irrigated fields, a nearly permanent solution to the problems of exposure to poor quality groundwater for drinking and the failure of drinking water wells in rural areas is found in central and north Gujarat.

Chapter 5 critically examines the role of irrigation management transfer in improving the performance of surface irrigation systems. These include a) delivery of timely and equitable amount of water to farmers; b) full irrigation revenue recoveries; and c) better operation and maintenance of irrigation systems. The authors' research shows that issues around the ineffectiveness of IMT are of governance and policy in nature. Thus, instead of attributing the failure in improving the performance of irrigation systems to the concept and practice of IMT, there is a need to revisit the policy formulation and implementation process that various states follow. Studies also suggest that the kind of institutional arrangements made and capacity building provided also impact on the success of IMT.

By drawing upon the experience of Gujarat, Madhya Pradesh (MP) and Maharashtra, the chapter authors are able to assess irrigation sector reforms and how changes have, or have not, been affected through the reform process. The selection of these three states presents a unique combination of IMT process undertaken in the country. In Gujarat, the process has been primarily facilitated through the efforts of civil society organizations, in MP it has been through the water resources department, and in Maharashtra both the civil society network and state irrigation department have been involved. Such comparative assessments explore the crucial link between the reform process, the efficacy of water users' associations (WUAs) and the eventual performance of the irrigation systems. Finally, suggestions are made, for the benefit of other developing countries, about sustainable institutional models for irrigation management transfer.

Along with growing evidence of water scarcity, the recent past has also seen a surge in governmental and non-governmental efforts to revive some of the lost traditional water bodies in the regions of poor water endowment. However, the role and benefits of water harvesting, especially in the context of naturally water scarce regions, is highly debated (Batchelor et al., 2002; Kumar et al., 2006; Ray and Bijarnia, 2006; Kumar et al., 2008d; Bassi and Kumar, 2010). In spite of the social, economic and environmental benefits that the traditional water bodies used to have, many had fallen in disuse over the years, by virtue of the advent of modern water supply systems which ensured higher reliability and better quality; thus, it becomes important to highlight instances where the traditional water harvesting systems have worked and the quantum of benefits they have been able to provide.

Chapter 6 presents an analysis of the case study undertaken in western Rajasthan, which is characterized by highly variable climate and has a history of traditional water harvesting. The main objective of this study was to evaluate the hydrological and socio-economic impacts of reviving khadins (traditional run-off farming systems) and traditional village ponds. It is important to mention here that these systems have been revived through the determined and sustained efforts of a philanthropic organization working in the region for the management of natural resources and community welfare. The study examines the hydrological impacts of these traditional systems by analyzing and comparing the water table fluctuations pre- and post-monsoon, in the areas both within and outside the influence area of the structures. It then undertakes a benefit–cost analysis of khadins, using an economic simulation of agricultural outputs that considers the wet and dry years and the active life of the system. Based on the findings of the study, the authors recommend factors to be considered in planning small water harvesting schemes under hydrological variability, in order to obtain optimum basin-wide benefits in sub-Saharan Africa.

It is now widely recognized that demand management in agriculture should be the strategy for improving groundwater balance in semi-arid and arid regions that are experiencing problems of depletion (Kumar, 2007, 2010). The introduction of water-efficient crops and micro-irrigation systems are widely advocated as the way to achieve water demand management in agriculture through crop water productivity improvements (Kumar, 2007; Kumar and Amarasinghe, 2009). Most of the past work which has analyzed the impact of MI considered plot and field as the unit, assuming that the cropping system remains the same and only the irrigation technology changes. While MI adoption is also associated with changes in cropping systems, with the high valued crops replacing traditional cereals and an expansion in the area under crops and irrigation, such an analysis will not provide a holistic assessment of the impact of the technology on overall farm income, food security and agriculture water use. Further, past research has not made a distinction between water saving at the field level and water saving at the farm level, nor between "applied water saving" and real water saving. Therefore, in order to capture the impacts on regional level water use in agriculture, groundwater sustainability and food security, the unit of analysis needs to change from plot and field to the farm (Kumar et al., 2008c).

The north Gujarat region is a clear example of the negative consequences of groundwater over-exploitation. For several decades, it has been undergoing significant changes in its farming systems as a result of several developmental interventions in the recent past. A project (the North Gujarat Groundwater Initiative), currently managed by the Society for Integrated Land and Water Management (SOFILWM), but launched by the International Water Management Institute with support from the Sir Ratan Tata Trust, Mumbai, has introduced water-efficient irrigation devices, water-efficient crops and land management practices among farmers in an effort to help them cut down groundwater use in irrigated agriculture without adversely affecting the economic prospects of farming. Several of the traditional crops such as wheat and bajra are being replaced by high valued crops, such as potato, groundnut, lemon, pomegranate and papaya. An estimated area of 73,000 acres of irrigated land is currently under MI systems in this region, including drips and sprinklers.

The effect of agricultural water management interventions at farming system has been the subject of a recent study. The main objectives of the research were to a) analyze the overall impact of the agricultural water management interventions on the farm income of the adopter households and aggregate groundwater use at the farm level; b) analyze the potential impact of a combinations of water demand management interventions at different scales of implementation on farm surpluses and regional groundwater use; and c) assess the implications of these for food security, and the risk and vulnerability of farming communities. Chapter 7 presents the findings of this study. The analysis makes a distinction between notional water saving, or applied water saving through the use of micro-irrigation systems, and real water saving. The key challenges in promoting micro-irrigation in sub-Saharan Africa are identified. It argues that, in the context of sub-Saharan Africa, MI systems will be economically viable only if the capital cost of the system is low, the cost saving in irrigation through a reduction in applied water and energy use is high, and the farmers adopt high value crops.

There are limits to which the economic value in the use of water can be enhanced when water is diverted for just one type of use. People need water for both domestic and productive needs, and the full value of water would be realized by them if they are able to use the water supply services for multiple uses. In the case of drinking water schemes, this would mean augmenting the quantity of water supplied through the system so that people are able to divert water for raising kitchen gardens, livestock drinking, etc. In the case of irrigation schemes, this means, people are able to get water of sufficient quality from a scheme near to their homes, with a sufficient degree of reliability to enable them to meet their domestic and livestock requirements.

Even if the agency has not designed the infrastructure for multiple uses, the system, by default, becomes a multiple use system. While some of the unplanned uses may get absorbed by the system, other uses can damage it (van Koppen et al., 2009). This is compounded by the often intermittent and unreliable nature of water supplies. Water supply systems that do not consider the livelihood needs of rural communities fail to play an important role in the latter's day to day life. Usually, the communities show only a low level of willingness to pay for the water supply services from such systems. This, in turn, affects the sustainability of the systems. Thus a vicious circle is perpetuated.

But there is growing appreciation of the fact that, whenever such unplanned uses take place from "single use systems" without causing much damage to the physical infrastructure, it brings about improvements in all four dimensions of livelihood related to water. These dimensions are freedom from drudgery; health; food production; and income (van Koppen et al., 2009). If households are able easily to access water from public irrigation schemes for their domestic needs and livestock uses, and if the reliability and quality of water are improved, then a lot of the investment in creating water supply infrastructure in the villages can be saved. But, as many scholars note, planning of water supply systems for multiple uses is restricted by a lack of comprehensive data on the incremental costs and returns (Meinzen-Dick, 2007; van Koppen et al., 2009).

In Chapter 8, we illustrate how the economic value of water from public irrigation systems, which are traditionally and dominantly used in irrigation, could be enhanced through converting them into multiple use systems. The authors refer in particular to the use of traditional tanks in south western Odisha; these serve multiple purposes, such as irrigation, cattle drinking, domestic water use and fisheries. The main objectives of this case study are to examine a) how physical factors such as climatic variability influence the water allocation priorities of tank users and the tanks' overall performance as a multiple use system; and b) how far physical systems improvements can increase their multiple use benefits for the poor.

The study shows that the tanks support fisheries, livestock water use and domestic water uses, in addition to irrigation, though irrigation is largest in terms of the size of the water economy. It also shows irrigation water usage suffers during normal rainfall years in spite of there being plenty of water in the tanks, while the economic value of outputs from the use of water was found to be quite significant during drought years, owing to the necessity of water for crop production. The reduced irrigation water use during normal rainfall years was due to reduced crop water demand during kharif season, the absence of a proper institutional mechanism for the efficient allocation of water amongst competing use sectors during the remaining part of the year, and a lack of sufficient infrastructure for water transport. It also shows how the economic value of outputs generated by the use of water could be enhanced both through a reallocation of monsoon water to irrigate high valued crops in winter and the introduction of technologies for taking water to distant commands.

Major lessons can be drawn, from the findings of the study, when designing public water systems for poor sub-Saharan African countries where government investment in formal rural drinking water supply and irrigation is still quite low. The ability of the poor rural communities to pay for water services in these countries is likely to be heavily influenced by whether the water supply services meet their livelihood needs along with domestic needs and, therefore, the scope for introducing the concept of multiple water uses while planning water systems is potentially high.

The current policies and programmes of the government for agricultural growth seem to focus on rain-fed areas, with the aim of enhancing productivity and production from such areas through water harvesting and the artificial recharge of groundwater. Huge investments under the National Rural Employment Guarantee Scheme (NREGS) have been made in the rural sector in order to generate employment. Most investment is for water-harvesting, drought proofing and water conservation. Many government-supported schemes are implemented in rural areas for groundwater recharge and watershed management without any scientific planning and technical supervision. The underlying assumption in much of this approach is that the growth in production and productivity in intensively irrigated areas of the country has declined, and that the future productivity growth has to essentially come from rain-fed areas. It is widely believed that water harvesting can support supplementary irrigation of rain-fed crops, and artificial recharge schemes can improve sustainability of well irrigation in the semi-arid and arid regions, which are largely rain-fed.

While the government of India continues to make investments in irrigation, no long-term strategies for large-scale irrigation development through investment in surface water schemes are envisaged. On the other hand, there is an over-emphasis on the role of groundwater recharge augmentation and small water harvesting schemes. While small water harvesting structures would provide reliable water supplies locally in certain hydrological settings, in most regions this would make little hydrological and economic sense. Adding to this, the policies governing the use of water in agriculture are degenerative, driven by political considerations, and promote inequity in access and inefficient use of what is a precious resource: they defeat the very purpose of sustainable agricultural production.

In Chapter 9, the authors discuss the present lopsided approach of the government and go on to argue that any programme and policies for agricultural growth in India should embark on a multi-pronged, region-specific strategy, which is rooted in the hydrological, agro-climatic and socio-economic realities of the regions concerned. The reality is that water-abundant regions have very little arable land that can be brought under irrigated production. They are dependent on food import from land-rich and water-scarce regions, which maintain high levels of production through over-appropriation of their limited surface and groundwater. Hence, the future irrigation in India lies in the appropriation of surface water in the water-abundant basins, and its export to, and use in, water-scarce river basins that are endowed with sufficient amount of arable land. This would help boost agricultural production, along with improving the sustainability of groundwater use.

In water-scarce regions, the policy emphasis should be on improving the productivity of water use in agriculture; hence, the focus should be on economic instruments such as water and energy pricing. However, too much reliance on this will also create problems, such as excessive preference for high value crops that use less water, with long-term negative consequences for food security and labour absorption in farming. In the long run, large-scale water imports would be required in the (agriculturally prosperous) semi-arid and arid regions, if we are to drought-proof these regions and sustain irrigated agriculture there. In water-abundant regions, the policies and programmes should be designed to encourage more intensive use of water.

A combination of technological, institutional and policy interventions helped Asian agriculture to undergo a major transformation in the 1960s, leading to the Green Revolution. But this sector is in crisis, perpetuated by declining productivity growth and average per capita holdings. Land fragmentation and water scarcity became severe: population growth, rapid urbanization and industrialization are some of the drivers. Paradoxically, the rural populations of most Asian countries depend primarily on agriculture for their livelihoods, in spite of the changing structure of the economy from agrarian to services. A major consequence of this will be the increasing vulnerability of the region's population to food insecurity and malnutrition. An important challenge for food security and agricultural production in the region is to attract investment in agriculture and irrigation, particularly in modernizing old irrigation systems and farms to improve the productivity of land and water. Chapter 11 explores the key areas of investment for promoting sustainable agriculture and food security, the strategic interventions in each area, the sources of investment and the key challenges.

Chapter 11 draws together key lessons from each chapter in order to provide developing economies with guidance in policy making in the field of water and agriculture. The chapter then focuses on sub-Saharan Africa, which is largely agrarian, and in which the irrigated area accounts for a small fraction of its cultivated area. Water security, as defined by the Water Poverty Index by Sullivan (2002) and the Sustainable Water Use Index by Kumar et al. (2008a), is one of the lowest in the region.

The analysis presented in this final chapter shows that the root cause of food insecurity in the region is poor water security. The key challenges in tackling it are discussed. It then goes on to describe the land and water management strategies for food security and sustainable agriculture in the region. The chapter also discusses some key areas for future research that would contribute to framing national policies for food security. One of the issues which needs to be investigated is whether groundwater can be the future of Africa's irrigation. Another important area for research is the current and future impact on food production and world cereal prices of an increasing allocation of arable land for bio-fuel production, and the resultant impact on food security. The third important area for research is to what degree low cost micro-irrigation systems are viable for the region.

Note

1. As per the projections, agriculture would be the major user of water in the year 2025 with 81.13 per cent. The domestic water requirement is expected to grow from 5.5 per cent in 2000 to 8.9 per cent in 2025 and industrial water requirement from a mere 2.0 per cent in 2000 to 8.83 per cent in 2025.

References

Amarasinghe, U. A., Shah, T., Turral, H., & Anand, B. K. (2007). *India's water future to 2025–2050: Business-as-usual scenario and deviations.* Research report 123. Colombo, Sri Lanka: International Water Management Institute.

Amarasinghe, U. A., Sharma, B. R., Aloysius, N., Scott, C. A., Smakhtin, V., de Fraiture, C., Sinha, A. K., & Shukla, A. K. (2005). *Spatial variation in water supply and demand across river basins of India.* Research report 83. Colombo, Sri Lanka: International Water Management Institute.

Banik, D. (1997). *Freedom from famine: The role of political freedom in famine prevention.* Dissertation and thesis 4/97. Oslo, Norway: Centre for Development and the Environment, University of Oslo.

Barrios, S., Berinelli, L., & Stroble, E. (2010). Trends in rainfall and economic growth in Africa: A neglected case study of the African growth tragedy. *The Review of Economics and Statistics, 92* (2), 350–366.

Bassi, N., & Kumar, M. D. (2010). *NREGA and rural water management in India: Improving the welfare effects.* Occasional paper 3. Hyderabad, India: Institute for Resource Analysis and Policy.

Batchelor, C., Singh, A., Rama, M. S., Rao, M., & Butterworth, J. (2002). *Mitigating the potential unintended impacts of water harvesting.* Paper presented at the IWRA international regional symposium on water for human survival, New Delhi, India, 26–29 November.

Cervantes-Godoy, D., & Dewbre, J. (2010). *Economic importance of agriculture for poverty reduction.* OECD food, agriculture and fisheries working papers no. 23. Paris, France: OECD Publishing.

Chaturvedi, M. C. (2000). Water for food and rural development: Developing countries. *Water International, 25* (1), 40–53.

D'Souza, D. (2002). *Narmada Dammed: An enquiry into the politics of development.* New Delhi, India: Penguin Books.

Datt, G. (1997). *Poverty in India and Indian states.* Washington, DC, USA: International Food Policy Research Institute.

Department for International Development (2004). *Climate change and poverty: Making development resilient to climate change.* London, UK: Department for International Development.

Dharmadhikary, S. (2005). *Unravelling Bhakra: Assessing the temple of resurgent India.* New Delhi, India: Manthan Adhyayan Kendra.

Evenson, R., Pray, C., & Rosegrant, M. (1999). *Agricultural research and productivity growth in India.* Research report 109. Washington, DC, USA: International Food Policy Research Institute.

Falkenmark, M., & Rockström, J. (2004). *Balancing water for humans and nature: the new approach in ecohydrology.* London, UK: Earthscan.

Government of India (1999). *Integrated water resource development: A plan for action. Report of the National Commission on Integrated Water Resources Development, vol. I.* New Delhi, India: Ministry of Water Resources, Government of India.

Government of India (2002). *India's planning experience.* New Delhi, India: Planning Commission, Government of India.

Government of India (2008). *Draft 11th five year plan.* New Delhi, India: Planning Commission, Government of India

Grey, D., & Sadoff, C. (2007). Sink or swim: Water security for growth and development. *Water Policy, 9* (6), 545–571.

Human Development Report (2006). *Beyond scarcity: Power, poverty and global water crisis.* New York, USA: United Nations Publishing.

International Water Management Institute (2007). *Water figures: Turning research into development.* Newsletter 3. Colombo, Sri Lanka: International Water Management Institute.

Kumar, M. D. (2003). *Food security and sustainable agriculture in India: The water management challenge.* Working paper 60. Colombo, Sri Lanka: International Water Management Institute.

Kumar, M. D. (2007). *Groundwater management in India: Physical, institutional and policy alternatives.* New Delhi, India: Sage Publications.

Kumar, M. D. (2010). *Managing water in river basins: Hydrology, economics and institutions.* New Delhi, India: Oxford University Press.

Kumar, M. D. (2011). Humanising water security. *The India Economy Review, 8* (2) 80–94.

Kumar, M. D., & Amarasinghe, U. A. (Eds.). (2009). *Water productivity improvements in Indian agriculture: potentials, constraints and prospects.* Strategic analysis of national river linking project of India series 4. Colombo, Sri Lanka: International Water Management Institute.

Kumar, M. D., & Singh, O. P. (2005). Virtual water in global food and water policy making: Is there a need for rethinking? *Water Resources Management, 19* (6) 759–789.

Kumar, M. D., Ghosh, S., Patel, A., Singh, O. P., & Ravindranath, R. (2006). Rainwater harvesting in India: Some critical issue for basin planning and research. *Land Use and Water Resources Research, 6* (1) 1–17.

Kumar, M. D., Malla, A. K., & Tripathy, S. (2008b). Economic value of water in agriculture: Comparative analysis of a water-scarce and a water-rich region in India. *Water International, 33* (2) 214–230.

Kumar, M. D., Patel, A., Ravindranath, R., & Singh, O. P. (2008d). Chasing a mirage: Water harvesting and artificial recharge in naturally water-scarce regions. *Economic and Political Weekly, 43* (35) 61–71.

Kumar, M. D., Shah, Z., Mukherjee, S., & Mudgerikar, A. (2008a). Water, human development and economic growth: Some international perspectives. In M. D. Kumar (Ed.), *Managing water in the face of growing scarcity, inequity and declining returns: Exploring fresh approaches* (pp. 842–858). Hyderabad, India: International Water Management Institute.

Kumar, M. D., Turral, H., Sharma, B. R., Amarasinghe, U. A., & Singh, O. P. (2008c). Water saving and yield enhancing micro irrigation technologies in India: When and where can they become best bet technologies? In M. D. Kumar (Ed.), *Managing water in the face of growing scarcity, inequity and declining returns: Exploring fresh approaches.* Hyderabad, India: International Water Management Institute.

Meinzen-Dick, R. S. (2007). Beyond panaceas in water institutions. *PNAS, 104* (39), 15200–15205.

Mukherjee, A. (2003). Groundwater development and agrarian change in eastern India. *IWMI-Tata Comment, 9* (2003), 11.

Mukherjee, S., Shah, Z., & Kumar, M. D. (2010). Sustaining urban water supplies in India: Increasing role of large reservoirs. *Water Resources Management, 24* (10), 2035–2055.

Ray, S., & Bijarnia, M. (2006). Upstream vs downstream: Groundwater management and rainwater harvesting. *Economic and Political Weekly, 41* (23), 2375–2383.

Schultz, T. (1979). The economics of being poor. *The Journal of Political Economy, 88* (4), 639–651.

Shah, T. (2001). *Wells and welfare in Ganga basin: Public policy and private initiative in eastern Uttar Pradesh, India.* Research report 54. Colombo, Sri Lanka: International Water Management Institute.

Shah, T. (2009). *Taming the anarchy: Groundwater governance in South Asia.* Washington, DC, USA: Resources for the Future.

Shah, Z., & Kumar, M. D. (2008). In the midst of the large dam controversy: Objectives, criteria for assessing large water storages in the developing world. *Water Resources Management, 22* (12), 1799–1824.

Sharma, K. D. (2009). Groundwater management for food security in India. *Current Science, 96* (11), 1444–1447.

Singh, O. P. (2004). *Water-intensity of north Gujarat's dairy industry: Why dairy industry should take a serious look at irrigation?* Paper presented at the second international conference of the Asia Pacific Association of Hydrology and Water Resources, Singapore, 5–9 June.

Singh, O. P., & M. Dinesh Kumar (2009). Impact of dairy farming on agricultural water productivity and irrigation water use. In M. D. Kumar & U. A. Amarasinghe (Eds.). *Water productivity improvements in Indian agriculture: Potentials, constraints and prospects.* Strategic analysis of national river linking project of India series 4. Colombo, Sri Lanka: International Water Management Institute.

Sullivan, C. (2002). Calculating water poverty index. *World Development, 30* (7), 1195–1211.

Suresh Kumar, D., & Palanisami, K. (2011). Can drip irrigation technology be socially beneficial? Evidence from southern India. *Water Policy, 13* (4), 571–587.

van Koppen, B., Moriarty, P., & Boelee, E. (2006). *Multiple use water services to advance the millennium development goals.* Research report 98. Colombo, Sri Lanka: International Water Management Institute.

van Koppen, B., Smits, S., Moriarty, P., Penning de Vries, F., Mikhail, M., & Boelee, E. (2009). *Climbing the water ladder: Multiple-use water services for poverty reduction.* Technical paper series 52. The Hague, The Netherlands: IRC International Water and Sanitation Centre.

2 Key issues in Indian irrigation

M. Dinesh Kumar, A. Narayanamoorthy and M. V. K. Sivamohan

Introduction

India has the world's second largest irrigated area; major irrigation projects over the years have contributed to expanding the irrigated area. A few scholars have recently documented the larger socio-economic (Bhalla and Mukherji, 2001) and welfare impact (Perry, 2001; Shah and Kumar, 2008) of large surface irrigation projects. Private well irrigation systems have, in the last three decades, witnessed rapid growth, surpassing that of flow irrigation in its contribution to the net irrigated area (Debroy and Shah, 2003; Kumar, 2007) by virtue of massive rural electrification, heavy electricity subsidies and institutional financing for pump sets (Kumar, 2007).

In recent years, a lopsided view, favouring only private well irrigation in preference to canal irrigation, has emerged among a few irrigation scholars in India (see, for instance, IWMI, 2007; Shah, 2009). The distorted thinking of considering one system superior to the other came because of a poor understanding of the determinants of irrigation growth, the fundamental difference between well and surface irrigation, and basic concepts in hydrology and water management. This has led to five misconceptions in the irrigation sector: a) that future growth in India's irrigation would come from groundwater (Amarasinghe et al., 2008); b) that well irrigation will have a big role in future agricultural growth and rural poverty alleviation in the water-abundant regions of eastern India (Shah, 2001; Mukherji, 2003; IWMI, 2007); c) that surface irrigation systems are highly inefficient; d) that, of late, returns on investments in surface irrigation systems have become negative; and e) local water harvesting and recharge can help sustain well irrigation in semi-arid and arid regions (Shah et al., 2003; Shah, 2009).

The key questions being investigated in this chapter are a) can well irrigation alone sustain expansion in India's irrigated area or, in India's water resource-water demand scenario, can canal irrigation be replaced by well irrigation? b) is surface irrigation really inefficient? c) does the declining area under canal irrigation mean negative returns on investments in surface irrigation systems? d) can local rainwater harvesting and recharge arrest groundwater depletion and sustain well irrigation economy? and e) can well irrigation boost agricultural growth and alleviate rural poverty in water-abundant east India?

Analyses, data type and sources

The sets of analysis used in this chapter are the per capita groundwater withdrawal in different states (m^3/capita/annum); the intensity of groundwater use in different states of India (m^3/m^2 of cultivated land); the per capita arable land in different states (ha/capita); the per capita effective renewable water availability per unit of arable land in selected basins of India (m^3/capita per annum); and the per capita agricultural water demand (m^3/capita/annum) in these river basins (8 of them).

The secondary data used for the analysis included district-wise utilizable groundwater resources and ground water draft (year 2005); the rate of siltation of major Indian reservoirs; the gross area irrigated by different sources in different states of India (year 2000); cultivable area available in major river basins of India (year 2000); the minor irrigation census data of 2001 for selected Indian states; the utilizable surface water resources of major Indian river basins; and the estimates of references evapo-transpiration in the upper and lower catchments of these basins, estimated using FAO's CROPWAT model. The secondary data were collected from a wide range of sources, namely the Central Ground Water Board, which is the premier scientific institution in India concerned with planning and evaluation of groundwater; the Central Water Commission, another scientific institution at the national level dealing with surface water resources; the Ministry of Agriculture; and a report of the National Commission on Integrated Water Resources Development.

In addition, we have extensively used analysis provided in several research papers in national and international publications, including those by the authors. Added to this are recent field visits conducted in different parts of rural India, including the command areas of large surface irrigation projects which are the subject of some ongoing research projects.

The future of India's irrigation: canals or wells?

Surface irrigation systems provide more dependable sources of water than groundwater-based systems in most parts of India.[1] For flow irrigation, there should be a dependable source of water and a (topography permitting) flow by gravity to the places of demand. Ideally, the design itself ensures yield from the catchment that is sufficient to supply water to the command areas for the designated duty, or, in other words, the design of the command area is adjusted to match the flows available from the catchment. Hence, the design life of the scheme will be more or less realistic for reliable "dependable yield" estimates, unless major changes occur in the catchment that alter the flow regimes and silt load.

But, in the case of groundwater, thousands of farmers dig wells drawing water from the same aquifer. Given the open access nature of the aquifer, all the agricultural pumpers of groundwater operate individually. Obviously, the "safe yield" of the aquifer is not reckoned while designing the well, so the productive life of a well is not in the hands of an individual farmer who owns it, but depends on the characteristics of the aquifer, wells and total abstraction. Two-thirds of India's geographical area is underlain by hard rock formations with poor groundwater potential (GOI, 2005). Most of peninsular and central India and some parts of western India are underlain by hard rock aquifers of basaltic and granitic origin.

In these areas, the highly weathered zone formations which yield water have small vertical depth – up to 30 metres. When the regional groundwater level drops below this zone, farmers are forced to dig bore wells, tapping the zone with poor weathering. These bore wells have poor yields, unlike the deep tube wells in alluvial areas such as north Gujarat, alluvial Punjab, Uttar Pradesh and Haryana. For instance, analysis of the census data provided in Table 2.1 shows that as many as 40 per cent of the 85,601 deep bore wells (that are in use) in AP were not able to utilize their potential due to poor discharge. The figure was nearly 19.1 per cent for Rajasthan, which has semi-consolidated and hard rock aquifers. The figure was 59.9 per cent for Maharashtra, which has basalt formations in nearly 95 per cent of its geographical area. Therefore, in spite of an explosion in well numbers, the well irrigated area has not increased in these areas during the past decade.

Second, growth rate in well irrigation has virtually decelerated in most parts of India since the 1990s. Table 2.2 shows that most well irrigation in India is concentrated in the arid and semi-arid regions of northern, north-western, western and peninsular India. In terms of per capita groundwater withdrawal per annum, intensive well irrigation is highest in some of the northern and north-western States, such as Punjab (1,729.9 m^3/capita/annum), Rajasthan, UP and Haryana and, to an extent, Gujarat, Tamil Nadu and Andhra Pradesh (Figure 2.1).

Intensive irrigation can be sustained for many decades in only a few pockets such as alluvial Punjab and Haryana and UP. This is because these regions are underlain by very good deep alluvial aquifers which are regionally extensive (GOI, 1999, 2005). These regions are already saturated in terms of irrigated area, and expansion in the irrigated area is not possible. In contrast, in Rajasthan, Gujarat, Andhra Pradesh and Tamil Nadu, problems of over-exploitation have halted further growth in well irrigation. Most of the untapped groundwater is in the eastern

Table 2.1 Percentage of dug wells and deep tube wells suffering from poor discharge in selected Indian states

Source no.	*Name of the state*	*Number and percentage of wells in use which face discharge constraints*	
		No. of deep tube wells	*Per cent of deep tube wells*
1	Andhra Pradesh	34216	40.0
2	Gujarat	20282	24.5
3	Madhya Pradesh	17841	58.5
4	Maharashtra	39958	59.9
5	Odisha	132	7.7
6	Punjab	10	0.10
7	Rajasthan	10010	19.1
8	Tamil Nadu	22838	34.1
9	Uttar Pradesh	3110	9.3

Source: authors' own analysis based on Minor Irrigation Census data, 2001.

Table 2.2 Gross irrigated area and well irrigated area for major Indian states

Sr. no.	*Name of the state*	*Gross irrigated area*	*Gross groundwater irrigated area*	*Percentage contribution of groundwater*
1	Andhra Pradesh	5.74	2.45	42.68
2	Bihar	4.55	2.43	53.50
3	Gujarat	3.51	2.81	80.06
4	Haryana	5.22	2.57	49.23
5	Karnataka	3.17	1.19	37.54
6	Madhya Pradesh	4.59	3.10	67.54
7	Maharashtra	3.82	2.63	68.85
8	Odisha	2.39	0.62	25.94
9	Punjab	7.80	5.92	75.90
10	Rajasthan	6.60	4.30	65.15
11	Tamil Nadu	3.50	1.88	53.71
12	Uttar Pradesh	17.67	13.42	75.95
13	West Bengal	3.50	2.13	60.86

Source: Government of India, Ministry of Agriculture, 2000.

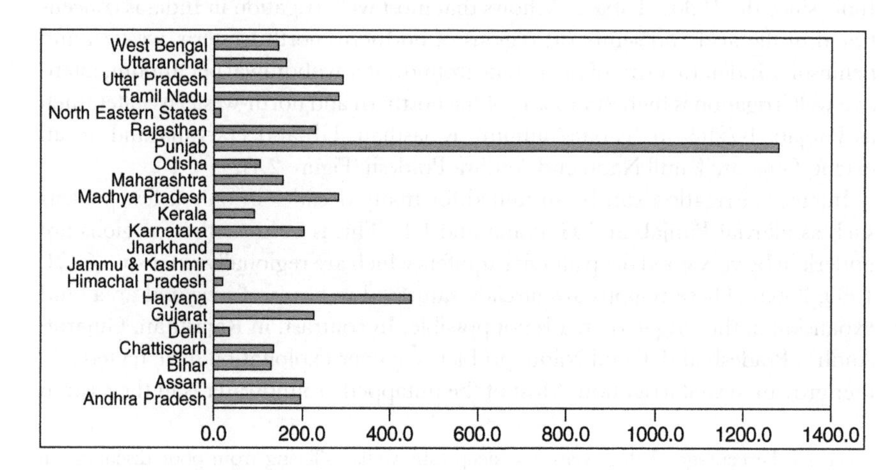

Figure 2.1 Per capita groundwater withdrawal rates in different states (m³/annum).

Source: Kumar et al., 2008b.

Gangetic plains, which are devoid of sufficient arable land (Kumar and Singh, 2005; Shah and Kumar, 2008; Kumar et al., 2008b). Peninsular and central India have a lot of un-irrigated land. Agriculture is prosperous in this part of the country, and demand for water is only going to grow, yet well irrigation is experiencing a "leveling off" and, sometimes, decline due to over-exploitation and monsoon failure.

Analysis of time series data on well irrigation from Andhra Pradesh illustrates this point. The data presented in Figure 2.2 show that the gross well irrigated area in the state peaked in 2000–01. But, by 2003–04, the irrigated area dropped by nearly 0.16 million ha. Since then, there has been only a minor growth in gross irrigated area, with the highest figure recorded in 2005–06. But what is interesting is the fact that this had no correlation with the growth in number of wells. There were only 1.4 million wells in the state in 2000–01, but together they irrigated 2.6 million ha of land in gross terms. The average area irrigated by a well was 1.8 ha, but this declined to a record low of 1.07 ha in 2003–04, with the growth in the number of wells by 0.86 million not getting translated into a growth in irrigated areas. This is a common phenomenon in hard rock areas, and occurs as a result of well interference. With an increase in number of wells, the influence area of a well increases and the available groundwater gets distributed among a larger number of wells (Kumar, 2007). The average area irrigated by a well recorded a minor improvement in 2005–06 (1.12 ha). This could be attributed to factors such as an increase in recharge from rainfall, a change in cropping pattern, and an increase in groundwater pumping in command areas, resulting from reduced surface water release from canals for irrigation (Kumar et al., 2011).

The figures show that, by 2000–01, the state had an optimum number of wells (1.4 million) to provide maximum irrigation. The only apparent benefit which was achieved through the increase in the number of wells was distributional equity, with more farmers getting direct access to groundwater for irrigation. This is not to say that distributional equity is not an important concern in irrigation. In fact, most of the farmers who are late entries in well irrigation are small and marginal farmers. But with such high density of wells in hard rock areas, the economic efficiency of groundwater abstraction becomes extremely low. This is evident from three important facts: a) the rate of well failures is becoming alarming; b) the area irrigated by a well has reduced to 1.07 ha, down by more than 40 per cent from the highest figure of 0.18 million ha experienced in 2000–01; and c) the power consumption for irrigating a unit area with groundwater had increased, from 3569.0 kWh in 1990–91 to 4222.7 in 2000–01, and 5225.9 kWh in 2003–04.[2] Much of this increase can also be attributed to replacement of diesel engines or energization of manually operated wells, but one would expect that electrification of wells had taken place by the time the groundwater irrigated area touched almost the peak

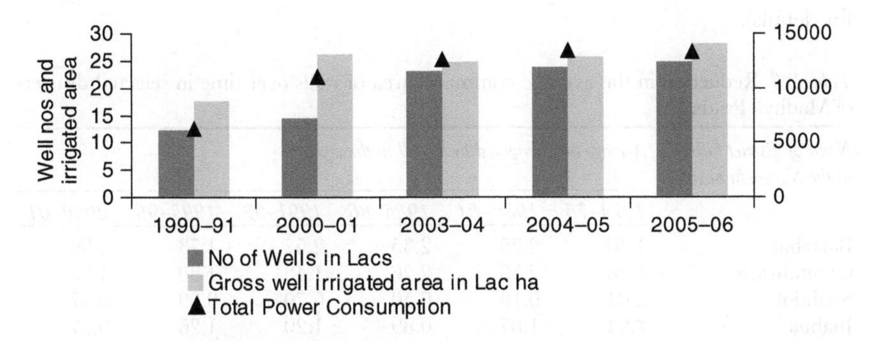

Figure 2.2 Well groundwater irrigated area nexus.

Source: based on data provided in Jain, 2008.

(i.e., in 2000–01). Therefore, it is safe to conclude that the increase in the energy requirement for groundwater pumping after 2000–01 was because of the increase in well numbers resulting from fast draw-downs in wells which, in turn, resulted from the phenomenon of well interference (Kumar et al., 2011). Well-interference is a characteristic feature of hard rock areas, which starts when all the groundwater that can be tapped is already tapped. In such situations, an increase in number of wells does not result in increase in total irrigated area (Kumar, 2007). Hence, it is wrong to assume that well irrigation in India could sustain the same pace of growth in coming years.

Another set of analysis carried out for the hard rock areas of the Narmada river basin in Madhya Pradesh showed that the average area irrigated by a single well has declined over a 25 year period (Table 2.3).

The spatial imbalance in resource availability and demand in India is aggravated by uneven distribution of surface water resources spatially. Nearly 69 per cent of India's surface water resources are in the GBM (Ganga-Brahmaputra-Meghna) basins (GOI, 1999). In the GBM basins, the water demand in agriculture is far less than the total renewable water resources, which is the sum of both renewable surface water and groundwater (Table 2.4), whereas in the five basins of south, western and Central India, the water demand for agriculture alone exceeds the renewable water resources (Table 2.5).

This imbalance can be effectively addressed only by large surface water projects, and not by groundwater projects.[3] Historically, water was taken from the rich upper catchments of river basins, which formed ideal locations for storage (Verghese, 1990). Surface irrigation can be expanded in future through investment in large reservoirs and transfer systems that can take water from the abundant regions of the north and east to the parched, but fertile, lands in the south, though their economic viability and social costs and benefits will have to be ascertained. It goes without saying that there is an influential lobby in India, as well as in neighbouring countries such as Bangladesh, which criticizes water management solutions which involve large-scale water transfer from north to south on scientific, technical, social, financial, ecological, economic and environmental grounds, though without much scientific data to support their arguments (see Kumar et al., 2008c for details).

Table 2.3 Reduction in the average command area of wells over time in selected districts of Madhya Pradesh

Name of district falling in the Narmada basin	*Average area irrigated by a well in these years*					
	1974–75	*1980–81*	*1985–86*	*1991–92*	*1995–96*	*2000–01*
Balaghat	4.50	2.25	2.35	2.57	1.73	1.96
Chhindwara	4.56	2.58	2.26	1.42	1.50	1.75
Shahdol	2.04	0.18	0.50	0.70	0.99	0.47
Jhabua	2.93	1.87	0.89	1.20	1.26	0.57
Betul	6.97	3.37	3.02	1.98	2.06	2.18

Source: Kumar, 2007.

Table 2.4 Per capita renewable water resources and per capita water demand in agriculture in two river basins

Sr. no.	*Name of the basin*	*Average annual rainfall in the basin (mm)*		*Average renewable water resources (m^3 / capita / annum)*	*Average effective water resources (m^3 / capita / annum)*	*Mean annual reference evapo-transpiration (mm)*		*Water demand for agriculture (m^3 / capita / annum)*
		Upper	*Lower*			*Upper catchment*	*Lower catchment*	
1	Ganga	1675.0	1449.0	1179.9	1399.4	710.0	1397.0	721.5
2	Brahmaputra	2359.0	2641.0	1737.1	2052.8	1064.0	1205.0	1180.9

Source: authors' own estimates based on ET_0 values estimated from FAO CROPWAT, and population, net and gross cropped area and renewable water availability figures obtained from GOI, 1999.

Table 2.5 Average reference evapo-transpiration against annual water resources in selected river basins in water-scarce regions

Sr. no.	*Name of the basin*	*Mean annual rainfall (mm)*		*Average annual water resources (mm)*	*Effective annual water resource (mm)*	*Reference evapo-transpiration (mm)*	
		Upper	*Lower*			*Upper*	*Lower*
1	Narmada basin	1352.00	792.00	444.70	937.60	1639.00	2127.00
2	Sabarmati basin	643.00	821.00	222.84	309.61	1263.00	1788.80
3	Cauvery basin	3283.00	1337.00	316.15	682.80	1586.90	1852.90
4	Pennar basin	900.00	567.00	193.90	467.80	1783.00	1888.00
5	Krishna basin	2100.00	1029.00	249.16	489.15	1637.00	1785.90

Source: Kumar et al., 2008.

Notes

a. The average renewable water resources (Table 2.4, column 5) of the basins were estimated by taking the sum of annual utilizable runoff (GOI, 1999, Table 3.6) and the dynamic groundwater resources from natural recharge in these basins (GOI, 1999, Table 3.9) and dividing it by the respective basin population. A considerable portion of the renewable water resources is un-utilizable because of the topography existing in these basins, and the peak flows. This un-utilized part can be treated as the flows available for ecosystems downstream after diversions.

b. The effective annual water resources of the basins were estimated by adding up the utilizable component of the renewable water resource and 1/100th of the static groundwater resources in the basin and then dividing it by the basin population. The total static groundwater resources in the two basins were estimated to be 7834.1 BCM and 917.2 BCM, respectively (GOI, 1999, Table 3.11, p. 46).

c. The water demand for agriculture (Table 2.4, column 9) of the basins was derived by multiplying the average of annual ET_0 from several locations, the per capita net cropped area in the basin, and the ratio of the gross cropped area to the maximum cropped area possible with 300 per cent cropping intensity. The net cropped area and gross cropped area figures considered for each basin are for 2050, as per the projections provided in the National Commission on Integrated Water Resources Development (GOI, 1999, Annex 3.2, p. 422). They are higher than the actual cultivated area in these basins at present. This leaves the chances of under-estimation of water demand for agriculture in our methodology.

But such transfer solutions literally do not hold much water, as the engineering feasibility of transferring groundwater in bulk is questionable. The greatest example is the hyper-arid north-western Rajasthan. The six districts of this region, which have endogenous surface water and saline groundwater, are now irrigated by water from the Indira Gandhi Canal, which carries water from Sutlej River in the Shivalik hills of Himachal Pradesh in north India. It irrigates a total of 2.035 million ha of land.

To what extent are surface irrigation systems inefficient?

Engineering efficiencies in large surface irrigation projects in India are much less than those of well irrigation schemes (GOI, 1999). Despite this, comparisons are used by some scholars to argue that surface irrigation projects are performing very badly, and that government investment in surface irrigation should be diverted to better management of aquifers (IWMI, 2007; Shah, 2009). While it goes without saying that management of canal irrigation leaves much to be desired, arguments about the bad performance of surface irrigation systems are based on obsolete irrigation management concepts which treated the water diverted from reservoirs in excess of crop water requirement as "waste" (Seckler, 1996; Howell, 2001). These comparisons do not reflect the economic efficiency of the entire system as the wastewater gets reused in the downstream part of the same system by well irrigators (Chakravorty and Umetsu, 2003).

As Seckler (1996) notes, the fundamental problem with this concept of water use efficiency based on supply is that it considers as inefficient both evaporative loss of water and drainage. It is not well informed by the water use hydrology of surface irrigation systems. Most of the seepage and deep percolation from flow irrigation systems replenishes groundwater, and is available for reuse by well owners in the command (Seckler, 1996; Allen et al., 1998). This recycling process not only makes many millions of wells productive, but also saves the scarce energy required to pump groundwater by lowering pumping depths. This is one reason why well irrigation can be sustained in many parts of Punjab and Haryana, Mulla command in Maharashtra, in the Krishna river delta of AP and Mahi and the Ukai-Kakrapar command in south Gujarat.

B. D. Dhawan, one of the renowned irrigation economists, looked at the economic returns from surface irrigation systems when he examined the merits of the claims and counter-claims about the benefits of big dams. He highlighted the social benefits generated by large irrigation schemes through the positive externalities, such as improvement in well yields, a reduction in incidence of well failures and the increased overall sustainability of well irrigation, citing the example of the Mulla command in Maharashtra (see Dhawan, 1990).

The social benefits (positive externalities) these canals generate by protecting groundwater ecosystems are immense (Shah and Kumar, 2008), reduced energy cost for pumping groundwater being one such benefit (Vyas, 2001). The likely impact of this on the energy economy of the country will be evident from the fact that electricity subsidy for agriculture in India was about INR 304.62 billion in

2001–02 (US $1 equals INR 50). Most of this goes to subsidizing pump irrigation in those states having large areas under well irrigation such as Punjab, Gujarat, Haryana, Madhya Pradesh, Maharashtra, Rajasthan, Tamil Nadu, Karnataka and Andhra Pradesh. The subsidy had increased from the previous nine years, from just INR 73.35 billion in 1992–93 (Planning Commission, 2002).

However, irrigation planners have nearly failed to capture these social benefits in the cost-benefit calculations (Shah and Kumar, 2008). Recent data from the government of Andhra Pradesh shows that the command irrigated regions of the state have the lowest number of groundwater "over-exploited" mandals. The tail end regions of the canals have a sufficient number of bore wells; these reap the benefit of return flows from canals and thus have good yields. The large reservoirs have raised cereal production to 42 million tonnes in the fifty years since independence. The social benefit this had generated by lowering cereal prices in the country has been estimated at INR 43 billion annually (Shah and Kumar, 2008). Added to these are the multiple use benefits that canal water generates, such as fish production and water for domestic and cattle use in rural areas.

Groundwater recharge using local runoff: a fallacy?

It is often suggested that flows from the small canals (*The Times of India*, 2008) or small water harvesting/artificial recharge structure (GOI, 2007; Shah, 2009) should be used for recharging aquifers. This is fallacious, as the arid and semi-arid regions, where aquifers are depleting (GOI, 2005; Kumar, 2007), have extremely limited surface water (Kumar et al., 2008a). Most of the over-exploited districts in India are in western and central Rajasthan; almost the entire Punjab; alluvial north Gujarat; and parts of Andhra Pradesh, Madhya Pradesh, Maharashtra and Tamil Nadu (GOI, 2005). The surface water resources in the basins of these districts are falling, are extremely limited and are already tapped using large and medium reservoirs (GOI, 1999). Hence, the over-exploited regions are also the regions of surface water shortage.

Any new interventions to impound water would reduce the downstream flows, creating a situation whereby Peter took Paul's water. Such indiscriminate water harvesting has also led to conflicts between upstream and downstream communities, as reported by Ray and Bijarnia (2006) in respect of Alwar in Rajasthan and Kumar et al. (2008a) in respect of Saurashtra in Gujarat. Kumar et al. show that in semi-arid and arid regions water harvesting/recharge not only has poor physical feasibility and economic viability, but has negative impacts on access equity in water (Kumar et al., 2008a).

The idea of dug well recharging for hard rock areas is being pushed, based on a false notion that it is a cheap (costing only INR 4,000 per well), easy and safe method of groundwater banking, unlike what is being practised in the United States and Australia (Shah et al., 2009). But this is far from the truth. Collecting runoff from the lowest points in the farm, channeling it to the well location, and then filtering it before finally putting in the well could be quite expensive, as land leveling and filter box construction costs may be significant, depending on the farm size and soil type.[4] Moreover, spending such large sums in no way guarantees environmental safety, as the runoff would contain fertilizer and pesticide, from the field, which cannot be removed using filters. Furthermore, as Kumar et al. (2008a) argue, the central government's INR 18 billion scheme to recharge groundwater through four million open wells in hard rock districts of the country, if implemented, would render many small and large reservoirs unproductive. Hence, well irrigation in peninsular and western India cannot be sustained unless water is brought from surplus basins in the east and north for recharging the aquifers there.

While policy makers, who had planned a large-scale initiative for groundwater recharge in rural areas, seem to be unaware of these undesirable consequences, what happens on the ground is even more deplorable. In most cases, the farmers take the benefit of this scheme using their land records and misappropriate the government funds, while field verification of the recharge structures is almost absent.

Bringing water from water-surplus basins to peninsular India would require large head works, huge lifts, long canals, intermediate storage systems, and intricate distribution networks. As we have argued, recharge schemes using local water are economically unviable. The reason is that, while the cost per cubic metre of recharge is abnormally high (see Table 2.6), the returns from irrigated crop production are far less (in the range of INR 1 to INR 17/m^3) as found in a study of the irrigation water productivity of various crops in nine agro-climatic sub-regions of the Narmada river basin in Central India (Kumar et al., 2008a). The need for vast precious land for spreading water for recharge would make it also socially unviable, while further increasing economic costs.

Since the aquifers in hard rock areas of India have extremely poor storage capacities, efficient recharge would require synchronized operation of recharge systems and irrigation wells. This would call for advanced hydraulic designs and sophisticated system operation. Therefore such an approach of using imported surface water for recharge would be akin to "catching the crane using butter". The fact that practising environmentally-sound artificial groundwater recharge in these water-scarce hard rock areas is a very expensive and complicated affair, requiring application of advanced science, is yet to be appreciated by a section of the water community.

Table 2.6 Estimated unit cost of artificial recharge structures built under CGWB pilot scheme

Sr. no.	*Type of recharge structure (life in years)*	*Expected active life of the system*	*Estimated recharge benefit (TCM)*	*Capital cost of the structure (in 100 thousand INR)*	*Cost of the structure per m^3 of water (INR/m^3)*	*Annualized cost* (INR/m^3)*
1	Percolation tank	10	2.0–225.0	1.55–71.00	20.0–193.0	2.00–19.30
2	Check dam	5	1.0–2100.0	1.50–1050.0	73.0–290.0	14.60–58.0
3	Recharge trench/shaft	3	1.0–1550.0	1.00–15.00	2.50–80.0	0.83–26.33
4	Sub-surface dyke	5	2.0–11.5	7.30–17.70	158–455.0	31.60–91.00

Source: Kumar et al., 2008a, based on GOI, 2007, Table 7, p. 14.

Hence, the best option would be for the farmers to use this expensive canal water for applying to the crops in that season of import (mainly monsoon season), and use the recharge from natural return flows for growing crops in next season. Opportunities for using water from "surplus basins" for recharging depleted aquifers exist at least in some areas, for example the alluvial north Gujarat and north-central Rajasthan. Ranade and Kumar (2004) have proposed the use of surplus water from the Sardar Sarovar Narmada reservoir during years of high rainfall to recharge the alluvial aquifers of north Gujarat through the designated command area in that region. They proposed the use of the existing Narmada Main Canal, and the rivers and ponds of north Gujarat for this, and their analysis showed that it is economically viable. It can protect groundwater ecology by reducing pumping; reduce the revenue losses in the form of electricity subsidy; and increase the flows in rivers that face environmental water scarcity, as well as giving direct income returns from irrigation. But Rath (2006) is absolutely correct that only crops having very high water use efficiency should be promoted in the commands receiving such water, so as to generate sufficient returns from irrigated production. However, this will be possible only if the price of irrigation water is pitched at such a level that it starts reflecting the scarcity value of the resource.

Is the contribution of surface irrigation declining?

In the year 2000, wells accounted for nearly 61 per cent of India's gross irrigated area (46.41 million ha), with the rest from canals and tanks (29.34 million ha) (GOI Agricultural Census, 2000). But to make a choice between surface scheme and groundwater scheme based on the crude numbers of "irrigated area by source" is "hydrologically and economically absurd". Which model of irrigation is best suited to the area in the future can be judged by the nature of its topography, hydrology and aquifer conditions. For instance, in rocky central and peninsular India, only imported surface water can sustain and expand well-irrigation. Indiscriminately embarking on well irrigation would only ruin the rural economy. Farmers in these regions desperately drill bore holes to tap water, doing so with high rates of failure (Kumar and Singh, 2008) and resultant farmer suicides. At least some scholars have begun to use "declining area under canal irrigation" to build a case for stopping investment in surface irrigation (*The Times of India*, 10 and 17 July, 2008, for Indian irrigation; Mukherji and Facon, 2009, for Asian irrigation; Shah, 2009). They seem to argue that the change in cumulative area irrigated by canals is a good indicator of the return on investment in surface irrigation systems. But this is a clear case of misuse of statistics. Such arguments come from poor understanding of how surface irrigation systems work. The fact is that, while completion of a new irrigation scheme increases the irrigated area of a particular locality or a region, this may not show up in the time series data on aggregate irrigated area at the national level. This phenomenon can be better understood if we look at the real factors that influence the irrigation performance of surface systems.

First, as a recent study in the Narmada river basin in central India shows, increased pumping of groundwater in upper catchments for agriculture can significantly reduce stream flows in basins where groundwater outflows contribute to surface flows (Kumar et al., 2006), thereby affecting the inflows into reservoirs. In fact, the whole country experienced a quantum jump in the number of agricultural wells during the 1970s and 1990s (Debroy and Shah, 2003). In a small watershed called Maheshwaram in Andhra Pradesh, with a drainage area of 64 sq. km (6400 ha), during 1975–2002 a total of 707 wells came up, all in the valley portions (National Geophysical Research Institute, 2002, as cited in Armstrong, 2004). Now, rights to groundwater are not clearly defined in India, and landowners enjoy the rights to use the groundwater underlying his or her piece of land (Saleth, 1996). Since agricultural wells are mostly private, de facto, groundwater is a private property. Therefore, it is beyond the institutional capacity of state irrigation bureaucracies to control such phenomenon occurring in the upper catchments of their reservoirs. Also, as is evident from the earlier discussions and from some studies, small water harvesting systems are adding to the reduction in inflows into reservoirs (Ray and Bijarnia, 2006; Kumar et al., 2008a).

Second, farmers in most surface irrigation commands install diesel pumps to lift water from the canals and irrigate the fields. Such instances are increasing, causing a pump "explosion" in rural India. Farmers can secure better control over water delivery through these pumps, and this is the reason for their preference for energy-intensive lifting to gravity flow. Another important reason is the illegal water diversion which is rampant in canal irrigation. The pumping devices enable illegal diversion of water for irrigating plots that are otherwise out of command due to topographical constraints. This was found to be rampant in many large irrigation commands, such as the Dharoi irrigation command in north Gujarat; the Mahi irrigation command in south-central Gujarat; and the Mulla-Mutha command in Maharashtra. The most recent example is the Sardar Sarovar Narmada project in Gujarat. Most of the delivery canals of this gravity irrigation scheme are contour canals. This means that only the lower side of the canals will have the command. However, extensive surveys carried out in the area showed that farmers whose land is located on the higher side lift the canal water were using diesel pump sets to irrigate their land. This is a widespread phenomenon. While such areas get counted as pump irrigated areas in government statistics, the direct water lifting from canals reduces that which can be brought under gravity irrigation.

Third, large reservoirs, primarily built for irrigation in this country, are being increasingly used for supplying water to big cities and small towns, as recent studies show. A recent analysis involving 301 cities/towns in India shows that, with an increase in city populations, the dependence on surface water resources for water supply increases to as much as 91 per cent for larger cities (Figure 2.3). Many large cities depend almost entirely on surface water imported from large reservoirs. Some examples are Bangalore, Ahmedabad, Chennai, Rajkot and Coimbatore, with contribution ranging from 91 to 100 per cent (ADB, 2007). Many of them depended in the past on local tanks, ponds and bore wells to meet their water needs.

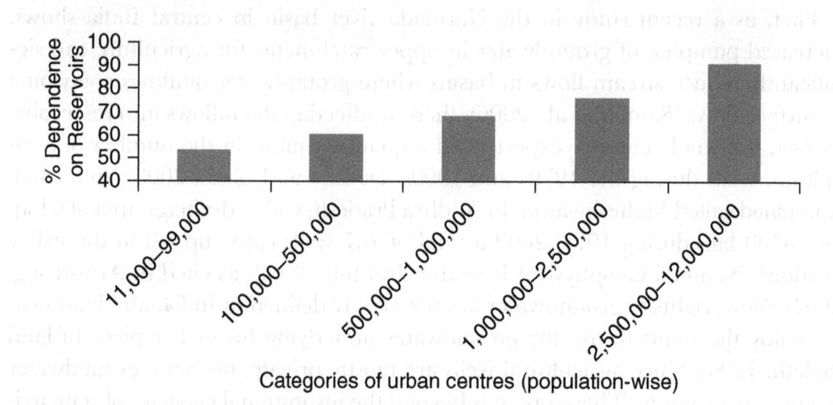

Figure 2.3 Degree of dependence of cities on reservoirs.
Source: Mukherjee and Shah, 2008.

Fourth, farmers in canal command areas, especially at the head reaches, tend to put more areas under water-intensive crops, ignoring the cropping pattern considered in the design. This is one of the reasons for shrinkage in the irrigated command area.

Fifth, water from large surface irrigation systems in many parts of India are used to feed tanks and ponds both in the command area, and also along the canal alignment, when the farmers do not need water at the time of its release in canals. This water is subsequently lifted using pumps to irrigate crops when the water release from canals stops. This gets counted as area irrigated by tanks/ponds and not as "canal irrigated area". These tanks/ponds also become ideal for raising fish and prawns, as found in the Godavari delta in Andhra Pradesh and the Mahi command in Gujarat.

Lastly, reservoirs are experiencing problems of sedimentation causing a reduction in their storage capacity and life, though worldwide experience has been that in some cases the rates are higher than those used at the time of design (Morris and Fan, 1998). The average annual loss of live storage for 23 large reservoirs in India with a total original live storage of 23,497 MCM (23.497 BCM) studied by the Central Water Commission was 213 MCM, i.e., an annual reduction of 0.91 per cent. Hence the loss of storage would be quite significant, particularly for older reservoirs. Such annual losses can sometimes reduce the effect of additions in storage achieved through new reservoir schemes on expanding irrigation.

Therefore, in the natural course of events and with the passage of time, the area under surface irrigation would decline, provided that nothing was done to revive the live storage of reservoirs. It is also, therefore, quite obvious that, with cumulative investment in surface irrigation systems going up with time, there may not be proportional rise in surface irrigated area. In order to evaluate the performance of surface schemes vis-à-vis return on investment, it is important to look at the performance of individual schemes in the light of these factors. At least some of those above facts are compelling reasons for fresh thinking on the planning and implementation of irrigation in India. Clearly, the solution does not lie in completely writing off surface systems in favour of wells as the latter are not a substitute for the former.

While these scholars lament the "dismal" performance of canal irrigation schemes in India, and stress the need to give impetus to well irrigation (Mukherji and Facon, 2009; Shah et al., 2009, p. 13), what is more noteworthy is the fact that the area under surface irrigation, which includes canal irrigation, tank irrigation and irrigation through canal and river lifting, has been steadily increasing during the past five-and-a-half decades and peaked in 2006–07, in spite of the myriad problems discussed above. Though there was a minor short term decline observed during 1993–94 and 2002–03, this decline was due to many factors. (Three of them are lack of adequate investments for new schemes (Planning Commission, 2008); droughts; and increasing diversion of water from reservoirs to urban areas.) In contrast, the growth in well irrigation declined significantly after 2000, with a growth rate of 0.18 million ha/year against 1.05 million ha/year observed during 1987–88 and 1999–00, and 0.634 million ha/year observed during 1967–68 and 1987–88 (see Figure 2.4). Sustaining well irrigation growth is a matter of concern, as 15 per cent (839) of the blocks/talukas/mandals in the country are over-exploited; 4 per cent are critically exploited and 10 per cent (550) are in the semi-critical stage (GOI, 2005), and these regions contribute very significantly to India's well irrigation.

Well irrigation for rural poverty reduction in eastern India?

Over the past few decades, well irrigation has been romanticized by some as a poverty-alleviating machine (Llamas, 2002; Debroy and Shah, 2003; Mukherjee, 2003; IWMI, 2007). While it is understood, and also well documented by many scholars in the past, that irrigation has a significant impact on poverty allevia-

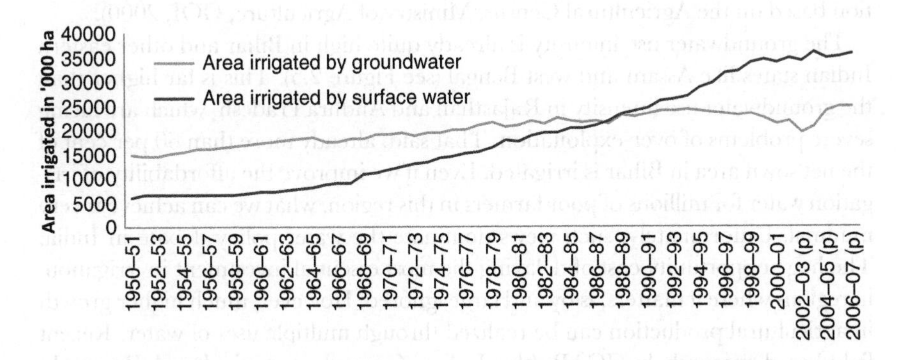

Figure 2.4 Net surface and well irrigated area in India (1950–51 to 2006–07).

Source: Based on data tables from Indiastat.

tion in rural areas (Bhattarai and Narayanamoorthy, 2003; Hussain and Hanjira, 2003), the over-emphasis on groundwater is somewhat difficult to understand. More strikingly, major arguments about the poverty impact of groundwater irrigation are made in the context of eastern India (Shah, 2001; Mukherji, 2003; IWMI, 2007). As Mukherji argues, "in regions of abundant rainfall and good alluvial aquifers, ground water irrigation can be a powerful catalyst in reducing poverty" (IWMI, 2007).

Eastern India's potential for triggering country-wide agricultural growth through a boost in well irrigation is also strongly argued (Shah, 2001; Mukherjee, 2003). Poor rural electrification and inadequate incentives for diesel pump dealers were blamed for the poor growth in well irrigation (Shah, 2001). Here, one really wonders about the actual effect of rainfall on irrigation demand and about the effect of irrigation versus land on economic surplus in areas of high water availability. Marginal returns from irrigation would be higher in areas of high aridity and low moisture availability, and not in humid/sub-humid areas with high moisture availability, as shown by an analysis which involved western Punjab and eastern Uttar Pradesh (Kumar et al., 2008c). Eastern India falls in the latter.

What is surprising is that, in the entire policy discourse on the impact of irrigation on agricultural development, the key factor of production – i.e., "land" – does not find a place anywhere. In fact, it is simply fallacious that a boom in well irrigation could be created in eastern India through proper rural electrification and energy policies. The reason is that water demand for irrigation is very low in this region.

The maximum water needed for irrigation is a direct function of both the per capita arable land and reference evapo-transpiration, and the inverse function of effective rainfall, provided the socio-economic conditions are favourable. In eastern India, not only is the rainfall high, but the ET is comparatively lower than western, north-western and southern India. The per capita arable land is lower than that of western, peninsular and north-western India (Kumar et al., 2008b). In Bihar, it is one of the lowest in the country with 0.068 ha against 0.17 ha in Punjab, and only 40 per cent of the net sown area remained un-irrigated in 2000 (information based on the Agricultural Census, Ministry of Agriculture, GOI, 2000).

The groundwater use intensity is already quite high in Bihar and other eastern Indian states like Assam and west Bengal (see Figure 2.5). This is far higher than the groundwater use intensity in Rajasthan and Andhra Pradesh, which are facing severe problems of over-exploitation. That said, already more than 60 per cent of the net sown area in Bihar is irrigated. Even if we improve the affordability of irrigation water for millions of poor farmers in this region, what we can achieve is very minimal. Unfortunately, such views dominate the water policy debate in India. The huge opportunity cost of delaying the most essential investment in irrigation, in regions where it matters, is by and large ignored. However, much higher growth in agricultural production can be realized through multiple uses of water. Recent field-based research by ICAR (the Indian Council of Agricultural Research) shows that well-designed multiple use systems can enhance the productivity of use of both land and water in eastern India remarkably. This involved integrating

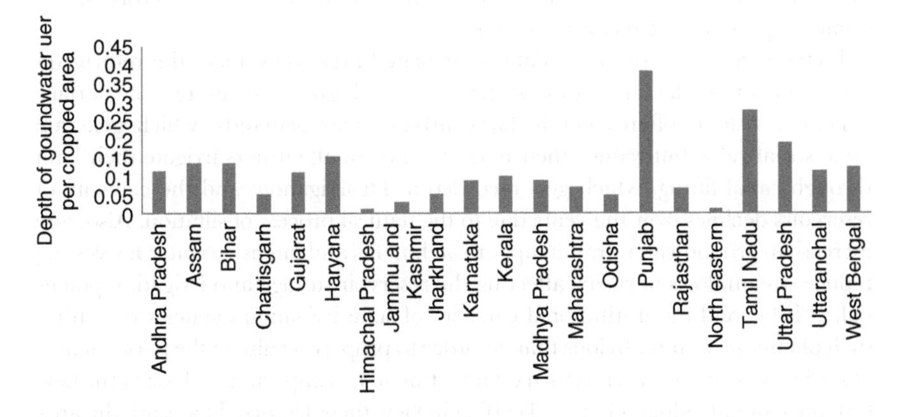

Figure 2.5 Intensity of groundwater use in different states.

Source: authors' own estimates based on GOI, 2000 and 2005.

fisheries, prawn farming and duckeries with paddy irrigation using local secondary reservoirs for the water (Sikka, 2009).

Conclusions and policy inferences

Evidence available from both the Indo-Gangetic plains and peninsular India suggests that there is a strong nexus between surface irrigation development and the sustainability of well irrigation. It is not prudent to invest in well irrigation without investment in large surface reservoirs and conveyance systems in semi-arid and arid areas. Risks associated with such irrigation development policies are more in the hard rock areas, as illustrated by the evidence from Andhra Pradesh and Madhya Pradesh. The spatial imbalance in water resource availability and water demand in India, which creates water-surplus regions and water-scarce regions, can be addressed only through surface water transfer projects.

The application of outdated irrigation management concepts leads to an under-appreciation of the benefits of surface irrigation. The positive externalities (social benefits) generated by surface irrigation, such as enhanced recharge of aquifers resulting from excessive return flows that sustain well irrigation; savings in the cost of energy used for pumping groundwater; and improved food security resulting from lowering of cereal prices, are missed out in the conventional cost-benefit calculations.

It is high time for the proponents of well irrigation to understand that water, whether well water or canal water, has to come from the same hydrological system. Promoting aquifer recharge using surface runoff from the same area to sustain well irrigation is hydrologically and economically absurd. The areas facing groundwater over-draft are experiencing extremely limited surface water resources, and artificial recharge schemes are economically unviable. A better appreciation of this fact would help save public funds, to the tune of thousands of crore, being spent on groundwater recharge schemes.

Using aggregate time series data on irrigated area to evaluate the returns on investments in surface irrigation systems will be highly misleading. The surface irrigation systems often cater to large urban water demands, which generate great social value but reduce their irrigation potential; farmers irrigate their land through canal lifting, which gets recorded as lift irrigation; and the capacity of reservoirs declines over the years due to the natural process of siltation. Also, any increase in groundwater draft in upper catchments and intensive water harvesting reduce stream flows in rivers, affecting the reservoir storage and irrigation potential. It is beyond the institutional capacity of state irrigation agencies to control such phenomena. It is obvious that, in order to properly evaluate the performance of surface schemes vis-à-vis return on investment, it is important to look at the performance of individual schemes, keeping in view these factors. That said, the area under surface irrigation has steadily increased during the past five decades.

The groundwater-abundant eastern India will not be capable of driving growth in well irrigation in future. A greater recognition of the fact that availability of arable land, rather than the availability of groundwater, is a major determinant of regional growth in irrigation demand would change the paradigm of water resource development for irrigation. The challenge is to build large water resource systems that are capable of transferring water from abundant basins to water-scarce basins having plenty of arable land, with minimum negative consequences for environment and ecology in both donor and receiving basins.

Groundwater in many semi-arid and arid areas suffers from poor quality owing to high mineral content and toxicity. They can pose new risks to crop production and food safety by affecting soils and plant tissues. To conclude, while problems facing canal irrigation are mostly managerial in nature, the problems are much more complex in the case of well irrigation, as both the physical and social science aspects of managing groundwater are much less advanced.

Notes

1 The authors here do not refer to reliability of water supplies, but dependability of the source of water, such as reservoirs, aquifer. It is understood that reliability of water supplies is better for groundwater-based irrigation schemes.
2 The consumption went down slightly in 2005–06. This could be due to improvements in groundwater conditions, in the form of rise in water levels, which actually can reduce the energy required for lifting a unit volume of groundwater.
3 This, however, does not to trivialize the role of demand management in regions where demand exceeds supplies.
4 Farmers in some parts of Gujarat, who have dried up open wells, laughed away this idea as impractical when it was mooted by two of the contributors of this article.

References

Allen, R. G., Willardson, L. S., & Frederiksen, H. (1998). Water use definitions and their use for assessing the impacts of water conservation. In J. M. de Jager, L. P. Vermes, & R.

Rageb (Eds.), *Proceedings of the ICID workshop on sustainable irrigation in areas of water scarcity and drought* (pp. 72–82). New Delhi, India: ICID.

Amarasinghe, U. A., Shah, T., & McCornick, P. (2008). Seeking calm waters: Exploring policy options for India's water future. *Natural Resource Forum, 32* (4), 305–315.

Armstrong, J. (2004). *International trends in water: A world and nation under stress.* Presentation made at the Water Summit 2004, Hyderabad, India.

Asian Development Bank. (2007). *Benchmarking and data book of water utilities in India.* New Delhi, India: Asian Development Bank and the Ministry of Urban Development, Government of India.

Bhalla, S., & Mukherji, A. (2001). Big dam development: Facts, figures and pending issues. *International Journal of Water Resources Development, 17* (1), 89–98.

Bhattarai, M., & Narayanamoorthy, A. (2003). Impact of irrigation on rural poverty in India: An aggregate panel data analysis. *Water Policy, 5* (5), 443–458.

Chakravorty, U., & Umetsu, C. (2003). Basinwide water management: A spatial model. *Journal of Environmental Economics and Management, 45* (1), 1–23.

DebRoy, A., & Shah, T. (2003). Socio-ecology of groundwater irrigation in India. In R. Llamas and E. Custodio (Eds.), *Intensive use of groundwater: Challenges and opportunities* (pp. 307–335). The Netherlands: Swets and Zetlinger Publishing.

Dhawan, B. D. (1990). *Big dams: Claims, counterclaims.* New Delhi, India: Commonwealth Publishers.

Government of India. (1999). *Integrated water resource development: A plan for action, Report of the national commission on water resources development, vol. 1.* New Delhi, India: Government of India.

Government of India. (2000). *Agricultural census 2000.* New Delhi, India: Government of India.

Government of India. (2001). *Minor irrigation census.* New Delhi, India: Ministry of Agriculture, Government of India.

Government of India. (2005). *Dynamic ground water resources of India.* New Delhi, India: Central Ground Water Board, Ministry of Water Resources, Government of India.

Government of India. (2007). *Report of the expert group on groundwater management and ownership.* New Delhi, India: Planning Commission, Government of India

Howell, T. (2001). Enhancing water use efficiency in irrigated agriculture. *Agronomy Journal, 93* (2), 281–289.

Hussain, I., & Hanjra, M. (2003). Does irrigation water matter for rural poverty alleviation? Evidence from South and South East Asia. *Water Policy, 5* (5), 429–442.

International Water Management Institute. (2007). *Water figures: Turning research into development.* Newsletter 3. Colombo, Sri Lanka: International Water Management Institute.

Jain, A. K. (2008). Challenges facing groundwater sector in Andhra Pradesh. Presentation by the Special Secretary to the Government of Andhra Pradesh.

Kumar, M. D. (2007). *Groundwater management in India: Physical, institutional and policy alternatives.* New Delhi, India: Sage Publications.

Kumar, M. D., & Singh, O. P. (2005). Virtual water in global food and water policy making: Is there a need for rethinking? *Water Resources Management, 19* (6), 759–789.

Kumar, M. D., & Singh, O. P. (2008). How serious are groundwater over-exploitation problems in India? Fresh investigations into an old issue. In M. D. Kumar (Ed.), *Managing water in the face of growing scarcity, inequity and declining returns: Exploring fresh approaches.* Hyderabad, India: International Water Management Institute.

Kumar, M. D., Ghosh, S., Singh, O. P., & Ravindranath, R. (2006). *Changing surface water-groundwater interactions in Narmada river basin, India: A case for trans-boundary water resources management*. Paper presented at the III international symposium on trans-boundary water management, UCLM, Ceudad Real, Spain, June.

Kumar, M. D., Malla, A. K., & Tripathy, S. (2008b). Economic value of water in agriculture: Comparative analysis of a water-scarce and a water-rich region in India. *Water International, 33* (2), 214–230.

Kumar, M. D., Patel, A., Ravindranath, R., & Singh, O. P. (2008a). Chasing a mirage: Water harvesting and artificial recharge in naturally water-scarce regions. *Economic and Political Weekly, 43* (35), 61–71.

Kumar, M. D., Sivamohan, M. V. K., & Narayanamoorthy, A. (2008c). *Irrigation water management for food security in India: The forgotten realities*. Paper presented at the international seminar on food crisis and environmental degradation: Lessons for India, New Delhi, India, 24 October.

Kumar, M. D., Sivamohan, M. V. K., Niranjan, V., & Bassi, N. (2011). *Groundwater management in Andhra Pradesh: Time to address real issues*. Occasional paper 4. Hyderabad, India: Institute for Resource Analysis and Policy.

Llamas, M. R. (2002). *Groundwater irrigation and poverty*. Keynote presentation at the international regional symposium on water for survival. Central Board of Irrigation and Power and Geographical Committee of International Water Resources Association, New Delhi, 27–30 November.

Morris, L. M., & Fan, J. (1998). *Reservoir sedimentation handbook: Design and management of dams, reservoirs and watershed for sustainable use*. New York, USA: McGraw Hill.

Mukherji, A. (2003). Groundwater development and agrarian change in eastern India. *IWMI-Tata Comment, 9* (2003), 11.

Mukherji, A., & Facon, T. (2009). *Revitalizing Asia's irrigation to sustainably meet tomorrow's food needs*. Colombo, Sri Lanka: International Water Management Institute and New York, USA: Food and Agriculture Organization of the United Nations.

Mukherjee, S., & Shah, Z. (2008). Large reservoirs: Are they the last oasis for survival of cities in India? In M. D. Kumar (Ed.), *Managing water in the face of growing scarcity, inequity and declining returns: Exploring fresh approaches*. Hyderabad, India: International Water Management Institute.

Perry, C. J. (2001). World commission on dams: Implications for food and irrigation. *Irrigation and Drainage, 50* (2), 101–107.

Planning Commission. (2002). *Annual report on working of state electricity boards and electricity departments*. New Delhi, India: Planning Commission, Government of India.

Planning Commission. (2008). *11th five year plan, 2007–2012: Agriculture, rural development, industry, services and infrastructure*. New Delhi, India: Planning Commission, Government of India.

Ranade, R., & Kumar, M. D. (2004). Narmada water for groundwater recharge in north Gujarat: Conjunctive management in large irrigation projects. *Economic and Political Weekly, 39* (31), 3510–3513.

Rath, N. (2005). India's water future: Responses to ITP's Delphi questionnaire. *IWMI-TATA Water Policy Research Highlights paper 16*.

Ray, S., & Bijarnia, M. (2006). Upstream vs downstream: Groundwater management and rainwater harvesting. *Economic and Political Weekly, 41* (23), 2375–2383.

Saleth, R. M. (1996). *Water institutions in India: Economics, law and policy*. New Delhi, India: Commonwealth Publishers.

Seckler, D. (1996). *The new era of water resources management: From dry to wet water saving*. Research report 1. Colombo, Sri Lanka: International Water Management Institute.

Shah, T. (2001). *Wells and welfare in Ganga basin: Public policy and private initiative in eastern*

Uttar Pradesh, India. Research report 54. Colombo, Sri Lanka: International Water Management Institute.

Shah, T. (2009). *Taming the anarchy: Groundwater governance in South Asia.* Washington, DC, USA: Resources for the Future.

Shah, T., Deb Roy, A., Qureshi, A., & Wang, Z. (2003). Sustaining Asia's groundwater boom: An overview of issues and evidences. *Natural Resources Forum, 27,* 130–141.

Shah, T., Kishore, A., & Hemant, P. (2009) Will the impact of 2009 drought be different from 2002? *Economic and Political Weekly, 44* (37), 11–14.

Shah, Z., & Kumar, M. D. (2008). In the midst of the large dam controversy: Objectives, criteria for assessing large water storages in the developing world. *Water Resources Management, 22* (12), 1799–1824.

Sikka, A. K. (2009). Water productivity of different agricultural systems. In M. D. Kumar and U. A. Amarasinghe (Eds.), *Water sector perspective plan for India: Potential contributions from water productivity improvements.* Colombo, Sri Lanka: International Water Management Institute.

Times of India, The. (2008). Wasting $50 billion dollars in major irrigation. *The Times of India,* 10 and 17 July.

Verghese, B. G. (1990). *Waters of hope.* New Delhi, India: Centre for Policy Research and IBH Publishing Co.

Vyas, J. (2001). Water and energy for development in Gujarat with special focus on the Sardar Sarovar Project. *International Journal of Water Resources Development, 17* (1), 37–54.

3 Food security challenges in India

Exploring the nexus between water, land and agricultural production

M. Dinesh Kumar, M. V. K. Sivamohan and A. Narayanamoorthy

Introduction

Food security had three major dimensions: food grain availability; access to food; and food absorption (MSSRF/WFP, 2010). Access to food, again, is a function of food prices. While prices are a complex interplay of several factors, such as supply shortfalls, drought- and flood-induced crop failures in major food grain exporting countries, and changing food consumption patterns (Chowdhury, 2011), for a country like India, whose cereal demand is one-sixth of the global demand, the prices would be heavily influenced by domestic production.

The per capita food grain availability in India has been declining alarmingly since 2001, with a major reduction in the production of cereals and pulses. On the other hand, the per capita demand for food grains is growing due to changing consumption patterns. While the average per capita demand for cereals for direct consumption has declined in the recent past, the demand for cereals for animal feed is growing owing to increased demand for animal products such as milk, beef, mutton, pork, poultry products, eggs and freshwater fish (Amarasinghe et al., 2007). The combined effect of rising per capita demand and declining per capita availability is food crisis, which manifested by the rising price of cereals and pulses in the market.

The total cereal demand in India is projected to be 291 million tons by 2025 (Amarasinghe et al., 2007). If these estimates are to be believed, food crisis is likely to intensify in future. There are three reasons for this. First, the maximum annual cereal production achieved so far in India has hardly touched 231 million tons (in 2007–08). Second, the net area under cultivation and area under food grains had more or less stagnated (GOI, 2008). Third, the growth rates in the yield of rice and wheat were much lower during the period 2001–02 to 2008–09 than during 1980–81 to 1989–90 (NRAA, 2011). In the event of future food crisis, expanding the area under cereals to enhance production would require additional water for irrigation to increase cropping intensities (Kumar, 2003). On the other hand, meeting the rising demand for milk and other dairy products would require intensive dairy production. Intensive dairy farming would be highly water-intensive in semi-arid and arid regions which are currently the major contributors to the country's dairy production. The reason is this would require more irrigated fodder crop (Singh, et al., 2004; Singh and Kumar, 2009) along with animal feed. In humid and sub-humid regions, it can result in nitrate pollution of shallow groundwater.

India's food and water crisis is often perceived to have been perpetuated by two factors. First is the widening gap between the utilizable water resources and the aggregate demand for water in agriculture and other sectors in certain regions. Second is the wide spatial and temporal variation in water resources endowments, in total contrast with the spatial and temporal variations in water demand, making certain regions water-abundant at certain times of the year (GOI, 1999, 2008). The food security challenge is being viewed by many as a challenge to manage, more economically, accessible water for increasing agricultural productivity and production in water-rich regions (Shah, 2001; IWMI, 2007; Sharma, 2009), and export of food from those regions to water-scarce regions. In the process, the constraints induced by the poor availability of arable land – let alone the issue of poor agricultural growth in those regions – have been, by and large, ignored.

Over the past three to four decades, widespread exploitation of groundwater has helped overcome the natural disadvantage India has, particularly because groundwater tapping can happen in arid and semi-arid regions which are poorly endowed in both surface and groundwater (Kumar, 2007; Sharma, 2009). But, this has, for a long time, shown a declining trend in many arid and semi-arid regions, which are also agriculturally prosperous, with an increasing number of blocks and districts falling in the "over-exploited" and "dark" categories (Kumar, 2007), while surface water resources in those regions have already been over-appropriated and over-allocated (Kumar, 2010). Yet some scholars believe that the continued exploitation of groundwater could sustain the boom in well irrigation in water-rich regions, and help avert the water crisis and growing food insecurity (Shah, 2001; IWMI, 2007; Sharma, 2009).

The purpose of this chapter is to address some of the misplaced notions of food security challenges posed by the country. To do this, we examine how far water and arable land become constraints in achieving food security in different regions; analyze the potential future impact of the availability of arable land and utilizable water resources, particularly groundwater, on the nation's food security; and suggest some broad strategies for achieving long-term sustainable water use and food security, while questioning some of the dominant theories relating to agricultural growth and regional food security.

Methods

In order to realize these objectives, we have first done the following. First, regional water balance scenarios are generated by comparing effective annual water resources against the water demand for agriculture in different regions. Then we examine the driving force behind intensive use of groundwater in different regions. Groundwater accounts for the lion's share of the agriculture production from irrigated areas in the country, and is the only resource which can be tapped to meet the demand–supply imbalances with the least financial, social and environmental consequences. We then consider the magnitude of groundwater over-exploitation problems and the way in which these can pose threat to agriculture production in different regions, and therefore threaten national food security. Various physical, social and economic and ethical considerations are discussed. Based on the estimates of groundwater contribution to irrigated area expansion in these regions, the degree of over-exploitation of groundwater there, and the respective regions' contribution to the production of major cereals in India, are assessed qualitatively, together with the implications of over-exploitation problems on national food security. Finally, some of the dominant theories concerning agricultural productivity and food security in the Indian context are weighted for their ability to address the problems of regional imbalances in groundwater development, agricultural growth and food production, using recent evidence from empirical research.

India's water supply and demand

From an anthropogenic perspective, water-scarce regions are those where the demand for water for various human uses far exceeds the total water available from the natural system, or where such water is available but the technology to access it is economically unviable. The total available water includes the surface water, water in the aquifers, and that held in the soil profile (Falkenmark, 2004). Water scarcity can also be felt when the resources are available in plenty in the natural system in a particular region, but adequate financial resources to access it are not available with the communities. The former is called physical scarcity, and the latter economic scarcity. In this article we are concerned with regions facing physical scarcity of water.

Physical scarcity of water occurs in "naturally water-scarce" regions or regions which experiences low to medium rainfalls and high evaporation rates (Kumar et al., 2006; Kumar et al., 2008b). Most parts of western, north-western, central and peninsular India fall under this category. They have low to medium rainfalls, and high potential evaporation rates. The mean annual rainfall ranges from less than 300 mm to 1000 mm, whereas the PE ranges from less than 1500 mm in some pockets in the north east to more than 3500 mm in some pockets in Gujarat and Maharashtra. In Chapter 2, we have explained the process which induces physical scarcity of water in the basins of naturally water-scarce regions using the illustration of Narmada, Sabarmati, Cauvery, Pennar and Krishna.

The "naturally water-rich regions" are those which experience medium to high rainfalls and low evaporation. Hence, eastern India, the eastern part of central India, the western Ghat region and the north-eastern region fall under this category. Their water demands are driven by the total amount of arable land, and the number of times with which it can be put to use in a year, rather than the total food demand. The reason is that there isn't much land available for utilizing the amount of water needed to produce this food, though water is available in plenty in these regions (Kumar, 2003). The analysis presented in Chapter 2 (on the future of India's irrigation – canals or wells), shows that the three major river basins (Ganga, Brahmaputra and Meghna) in the naturally water-rich regions are surplus basins.

Drivers of groundwater-intensive use

We have already seen that there is major mismatch between water supply and water demand for agriculture in India. Eastern India, extending over Bihar and eastern UP, which is part of the Gangetic alluvium, is abundant in both surface water and groundwater. This region is underlain by one of the richest aquifers in the world, having huge static groundwater reserves (GOI, 1999, Table 3.11, p. 46). Still, this region continues to be a net importer of food grain (Amarasinghe et al., 2004), and is agriculturally very backward (Evenson et al., 1999). The productivity levels for main cereals such as wheat and paddy are lowest in this region (NRAA, 2011).

There is scope for improving the productivity of main cereal crops such as wheat and paddy, which are major crops in this region, through enhancing farmers' access to well irrigation by means of massive electrification and pump subsidies. For example, in UP, which has the largest area under wheat, yields could be increased by 50 per cent; in Bihar, by over 100 per cent. Similarly, rice yields in Chattisgarh could be raised 150 per cent on un-irrigated land and 169 per cent on irrigated land (Planning Commission, 2007). But there are limitations on the extent to which this can contribute to enhancing the food grain production in the country. This limit mainly comes from poor land availability due to very high pressure on land; very little additional land that can be brought under irrigation; a high degree of land fragmentation; poor public investment in rural infrastructure including irrigation and electricity; ecological constraints due to floods; and overall lack of institutional and policy reforms in the agricultural sector.

As estimates of the productivity of food grains suggest (NRAA, 2011), the returns on irrigation also appear to be low for these sub-tropical water abundant regions, as compared to semi-arid tropics (Figure 3.1). Sub-tropical Bihar, with 60.7 per cent of the total cultivated area of food crops under irrigation has an average grain yield

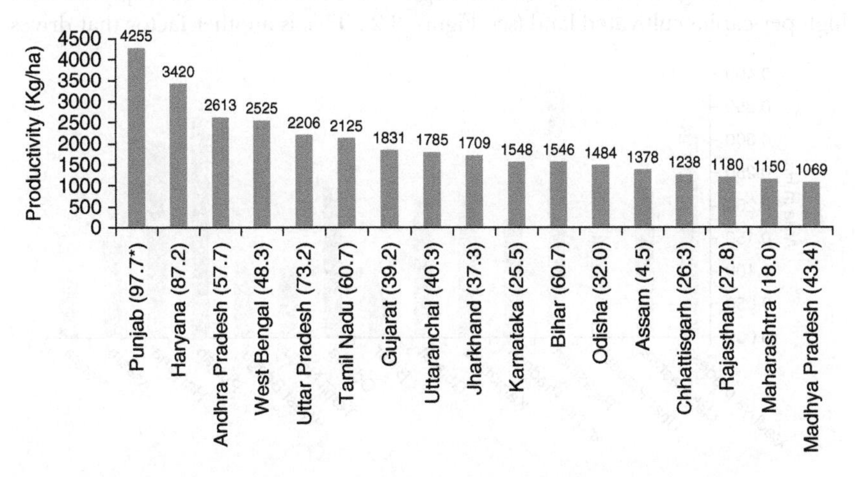

Figure 3.1 Production of food grains vis-à-vis irrigation in Indian states (2007–08).

Source: NRAA, 2011.

of 1546 kg/ha. It's a similar picture in Assam with an average yield of 1378 kg/ha, but Andhra Pradesh, in the semi-arid tropic, with a slightly lower percentage of food crops under irrigation (57.7 per cent) has an average grain yield of 2613 kg/ha.

Very small size holdings and low crop yields reduce the capacity of farmers to generate surpluses, and use these surpluses for investing in high yielding seed varieties and irrigation that can support these varieties. Very high land fragmentation forces farmers to depend on water buyers rather than investing in their own irrigation infrastructure, which would be economically inefficient due to poor utilization of the potential created (Kishore, 2004). With very low level of electrification, water buyers pay prohibitive prices for the water which is purchased from well owners, reducing the net returns from farming (Kumar, 2007). This is a possible reason for the low total factor productivity (TFP) growth in this region (Evenson et al., 1999).

On the other hand, the farmers in the semi-arid and arid regions of Punjab, Haryana, Gujarat, Andhra Pradesh and Tamil Nadu have been rather quick in adopting green revolution technologies, with modern high yielding varieties and farm mechanization, as large public investment in irrigation infrastructure supported this. The availability of sufficient amount of arable land enabled the farmers the quicker adoption of modern high yielding varieties, as they could produce enough surpluses from irrigating them. The subsequent years witnessed a rapid growth in wells and well irrigation, with the traditional varieties being replaced by modern high yielding varieties even in the non-command areas. Rapid rural electrification, followed by heavily subsidized electricity for groundwater pumping and institutional financing for wells and pump sets helped sustain intensive irrigation of water-intensive crops. This has led to over-exploitation of groundwater in these regions.

The much lower per capita net cultivated area and the lowest productivity levels for cereals such as wheat and paddy essentially means that the water-rich regions have severe food shortages, making these regions to depend on imports from the water-scarce regions that have forward agriculture, with both high crop yields and high per capita cultivated land (see Figure 3.2). This is another factor that drives

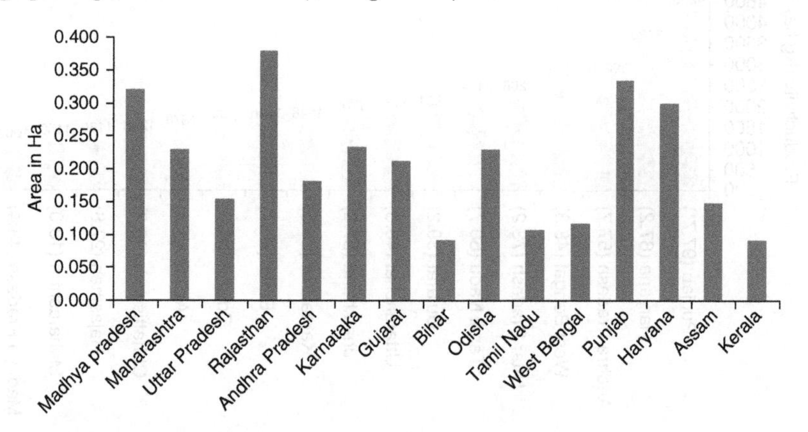

Figure 3.2 Per capita cropping in different states.

Source: authors' own estimates based on GOI, 2000 and 2001.

intensive well irrigation in the water-scarce regions, as the regional imbalances in food production and the concern of national food self-sufficiency ensure market support for the food produced in these regions through good procurement prices.

Assessing the magnitude of groundwater exploitation problems in India

There has been a lot of whistle-blowing about the impending groundwater crisis in many arid and semi-arid regions based on anecdotal evidence from some of these regions on groundwater level trends (Kumar, 2007). However, if one goes by the official estimates of groundwater development, only 23.1 million hectare metres out of the 43.2 million hectare metres of renewable groundwater in the country is currently utilized (GOI, 2005). Again, going by the disaggregated data, only 15 per cent of the groundwater basins in the country are over-exploited; 7 per cent critically exploited. Nearly 62 per cent of the groundwater basins are still "safe" for further exploitation (GOI, 2005). Interestingly, as per the official statistics, Punjab is one of the states where over-exploitation is most serious, next only to Rajasthan, and is followed by Delhi and Gujarat. But, as discussed in Chapter 2, the number of over-exploited districts in the hard rock areas of Andhra Pradesh, Tamil Nadu and Saurashtra in Gujarat, where high incidence of well failures is reported, is very low.

Therefore, such doomsday prophecies have not been based on rational assessment of the scenario using data on hydrological changes and hydrodynamics. This is not to say that groundwater over-exploitation is not a cause for concern in India. In the subsequent section, we examine how far these "doomsday prophecies" are correct.

Water level trend analysis

Groundwater level trends are a net effect of several changes taking place in the resource conditions caused by recharge from precipitation, return flows from irrigated fields, seepage from water carriers, abstraction or groundwater draft and lateral flows or outflows into the natural streams (Todd, 2003, pp. 218–229). In a region where long-term levels of groundwater pumping are less than the average annual recharge, the groundwater levels can experience short-term declining trends as a result of drastic increase in groundwater pumping owing to monsoon failure. Such a phenomenon does not represent the long-term trend. It is important to note here that semi-arid regions in our country also experience significant inter-annual variability in rainfall (based on Pisharoty, 1990; Kumar et al., 2006).

Furthermore, not all changes in groundwater conditions can be attributed to hydrological stress induced by human action. Natural factors can also cause hydrological stresses, as illustrated through a study of surface water and groundwater interactions in the Narmada river basin in India, where the change in groundwater outflows into streams was found to be an important determinant of the water level trends (Kumar, 2010). In such cases, maintaining such natural processes would limit the safe abstraction rates to levels much lower than that which is permissible on the basis of renewable recharge. In such situations, estimating the base flows would be crucial in arriving at the net utilizable recharge. Since groundwater outflows are not properly accounted for in the estimates of the net recharge, the estimates would show a much lower stage of development than that which the region is experiencing (Kumar and Singh, 2008).

Analyzing the groundwater balance in order to assess over-draft

Ideally, in a region where lateral flows and outflows from groundwater systems are insignificant, groundwater "over-draft" can take place if the total evapotranspirative demand for water (ET) per unit area is more than the total effective rainfall, i.e., the portion of the rainfall remaining in situ after runoff losses and the amount of water imported from outside for unit area. In many semi-arid to arid regions of India, cropping places an intensive demand on irrigation water during the winter and summer months. The ET demands for crop are much higher in comparison to the effective rainfall. The deficit has to be met either from local or imported surface water or groundwater pumping. Hence, the change in groundwater storage would be the imbalance between the total of recharge from rainfall and return flows from irrigation, and groundwater draft. In semi-arid and arid regions, natural recharge from precipitation are generally very low. In an area with intensive surface irrigation, a negative balance in groundwater indicates high levels of over-draft or deficit in effective rainfall in meeting the ET requirements (Kumar, 2007; Kumar and Singh, 2008).

Understanding geological characteristics

Determining under what geological conditions drops in water levels occur is also important in assessing the extent of groundwater over-draft conditions. Many semi-arid and arid areas in the country are hard rock areas. In these regions, the specific yield of aquifers is very small: 0.01 to 0.03. Large seasonal drops in water levels are a widespread phenomenon in these areas. During the monsoon, a sharp rise in water levels is observed and, after the monsoon, water levels start receding, so a clear distinction between seasonal depletion and annual depletion is to be made. Further, in hard rock areas, a unit volume of groundwater pumped from the aquifer results in up to 12 to 13 times the annual drawdown that occurs in alluvial areas for the same amount of over-draft. A fall in the water level of one metre in alluvial Punjab should be a cause for much greater concern than a one metre fall in the water levels in the hard rock areas of Tamil Nadu, Saurashtra or Karnataka, given the fact that the specific yield of alluvium in Punjab is in the range of 0.13–0.20. This will be evident from the data on the recharge abstraction balance for two distinct regions. This is not to say that the magnitude of water level drop is not important. A sharp fall in water level would also have serious implications for the investment required for pumping groundwater, and also the efficiency with which groundwater could be abstracted. What is more important is the long-term rate of decline in water levels.

Integrating the negative consequences of over-exploitation

The concept of "groundwater over-development" or aquifer over-exploitation is complex, and linked to various "undesirable consequences" which are physical, social, economic, ecological, environmental and ethical in nature (Custodio, 2000). Therefore, an assessment of groundwater over-development involves considerations that are hydrological, hydro-dynamic, economic, social and ethical in nature. However, some of the most important factors are the groundwater stock available in a region; water level trends; net groundwater outflows against inflows; the economics of groundwater-intensive use (particularly irrigation which takes lion's share of the groundwater in most semi-arid and arid areas); the criticality of groundwater in the regional hydro-ecological regime; and the ethical aspects and social impacts of groundwater use. Let us examine how the use of these in assessing groundwater over-draft would change the groundwater scenario in India.

As regards the groundwater stock, a region with huge amount of static groundwater resources, like the alluvial plains of Ganges, may experience over-draft conditions, with resultant steady decline in water levels (source: based on GOI, 1999). In such regions, assessing over-draft conditions purely in terms of average annual pumping and recharge may not make sense. The long-term sustainability goal in groundwater use can be realized even if one decides to deplete a certain portion of the static groundwater resources, along with the renewable portion, annually (Custodio, 2000). Limiting groundwater use to renewable resources, with the aim of benefiting future generations, can mean foregoing large present benefits.

As regards the influence of water level trends, a region may not experience over-draft when pumping is compared against recharge. Still, partial well failures could be an area of concern due to the seasonal drops in water levels. Such steep seasonal drops in water levels are characteristic of hard rock areas (Kumar et al., 2001).

As per official estimates, many such regions are still categorized as "white" and "grey", though these areas face severe groundwater scarcity during the summer (Kumar et al., 2001; Kumar and Singh, 2008). Table 3.1 shows the data on wells which have failed and those which are not in use, and is drawn from the 2001 Minor Irrigation Census of twelve Indian states. The total number of failed wells include both wells which have permanently gone dry and wells which are temporarily not in use. The latter category essentially refers to wells which are seasonal due to seasonal depletion of groundwater. The data shows that, in the states which are mostly underlain by hard rock formations, both the percentage of wells that have failed and those which are not in use are high. For instance, in Odisha, even as per 2005 official data, the stage of groundwater development was only 18 per cent (GOI, 2005). In spite of this, a large percentage of dug wells (21.5 per cent), and a much large percentage of deep tube wells (51.8 per cent) have failed. In terms of numbers, a total of more than 79,518 dug wells had failed in Odisha by 2001. A similar trend is found in Andhra Pradesh, Tamil Nadu and Madhya Pradesh.

Table 3.1 Well failures in different categories from eight major Indian states (2001)

Sr. no.	*Name of the state*	*Percentage of wells which have failed/not in use*		
		Dug wells	*Shallow tube wells*	*Deep tube wells*
1	Andhra Pradesh	17.3/20.20	2.4/2.9	1.6/2.2
2	Bihar	18.0/32.50	2.7/4.8	36.7/44.9
3	Gujarat	19.3/22.0	12.0/14.2	8.5/12.0
4	Madhya Pradesh	16.2/18.0	14.7/15.1	13.9/16.2
5	Maharashtra	9.30/10.9	4.3/7.9	10.7/13.6
6	Odisha	21.0/25.0	16.5/19.3	51.8/62.8
7	Punjab	0.0/0.0	0.0/0.0	1.2/1.6
8	Rajasthan	24.9/27.9	3.3/3.5	7.4/7.8
9	Tamil Nadu	20.0/22.1	7.5/8.1	19.7/20.4
10	Uttar Pradesh	4.4/9.50	0.80/1.2	3.7/5.0
11	West Bengal	6.30/10.3	3.5/4.4	9.8/12.2

Source: Kumar and Singh, 2008.

Note

a. The figures after the forward-slash (/) show the percentage of wells which are, for a variety of reasons, currently not in use.

Similarly, the current district-wise assessment of groundwater development does not take into account the long-term trends, as the latest methodology suggests. A region might have experienced a long-term decline or rise in water levels; however, a few years of abnormal precipitation may change the trends in the short term.

Another dimension of groundwater over-exploitation is economic. The cost of production of water should not exceed either the benefits derived from its use or the cost of provision of water from alternative sources. Drops in water levels beyond certain limit cause negative economic consequences, by raising the cost of abstraction, per unit volume of water, not only in irrigation but also in other sectors like municipal uses. Though there could be plenty of water in the aquifers, the fixed cost and variable costs of abstraction of water could be prohibitively high. An analysis of the Sabarmati river basin of north-central Gujarat, which experiences over-exploitation, shows that groundwater irrigation for the prevailing cropping system would be economically unviable if the farmers had to bear the full cost of the energy used for pumping water (Kumar et al., 2001).

In many areas underlain by basalt and granite, those highly weathered zones in the geological formations which yield water, have only a small vertical extent of up to 30 m. When the regional groundwater level drops below this, farmers are forced to dig bore wells; tapping groundwater from strata below this depth using open wells would be not only technically infeasible, but also economically unviable. These bore wells have poor yields. For instance, analysis of census data (Table 3.2) show that as many as 40 per cent of the nearly 85,601 deep bore wells (that are in use) in AP were not able to utilize their potential due to poor discharge. The figure was 19.1 per cent for Rajasthan, which had sedimentary and hard rock aquifers and 59.9 per cent for Maharashtra, which has basalt formations.

Table 3.2 Percentage of dug wells and deep tube wells suffering from poor discharge in selected Indian states

Sr. no.	*Name of the state*	*No. and percentage of wells in use which face discharge constraints*	
		No. of deep tube wells	*% of deep tube wells*
1	Andhra Pradesh	34216	40.0
2	Bihar	430	12.6
3	Gujarat	20282	24.5
4	Madhya Pradesh	17841	58.5
5	Maharashtra	39958	59.9
6	Odisha	132	7.7
7	Punjab	10	0.10
8	Rajasthan	10010	19.1
9	Tamil Nadu	22838	34.1
10	Uttar Pradesh	3110	9.3
11	West Bengal	15	0.30

Source: authors' own analysis based on the Minor Irrigation Census 2001 data.

Withdrawal of groundwater from these bore wells creates excessive draw-downs; and it is this, as well as high well interference, that causes well failures to become widespread. Therefore, before a farmer hits water in a successful bore well, he or she would have sunk money in many failed bore wells. Because of this, the actual cost of abstraction of groundwater becomes very high. The command area of wells is also on the downward trend, as shown by data from Madhya Pradesh (see Table 3.3). In Betul district, the average area irrigated by a well reduced from 6.97 ha to 2.18 ha over a 26 year period. So investment for well construction, compounded by a reduction in command area, reduces the overall economics of well irrigation. However, this aspect has been captured in the criteria for assessment of over-exploitation. As per the official data, these five districts are still in the "white" category, and safe for further exploitation (GOI, 2005).

The economics of groundwater use is not static. Economic viability of groundwater abstraction can change under two circumstances: a) opportunities for using the pumped water for more productive uses emerge with changing times; and

Table 3.3 Reduction in the average command area of wells over time in the Narmada basin, Madhya Pradesh

Name of district falling in the Narmada basin	*Average area irrigated by a well in ha*					
	1974–75	*1980–81*	*1985–86*	*1991–92*	*1995–96*	*2000–01*
Balaghat	4.50	2.25	2.35	2.57	1.73	1.96
Chhindwara	4.56	2.58	2.26	1.42	1.50	1.75
Shahdol	2.04	0.18	0.50	0.70	0.99	0.47
Jhabua	2.93	1.87	0.89	1.20	1.26	0.57
Betul	6.97	3.37	3.02	1.98	2.06	2.18

Source: authors' own estimates based on primary data as provided in Kumar, 2007.

b) the cost of abstraction of groundwater changes due to improvements in pumping technologies, or changes in the cost of the energy required for pumping groundwater. With massive rural electrification, the cost of groundwater abstraction in Bihar could come down to negligibly low levels. On the other hand, adoption of new high yielding varieties or high valued crops can increase the gross returns from farming.

The social consequences of groundwater use are equally important. One serious issue associated with groundwater-intensive use is that it excludes resource-poor farmers from directly accessing the resource when water levels start falling. Equity in access to resource should be an important consideration in assessing the degree of over-exploitation. In many areas, it is only the rich farmers who are able to pump groundwater, owing to an astronomical rise in the cost of drilling wells, and they enjoy unlimited access to the resource. While the well owners of Mehsana incur an implicit cost of nearly INR 0.5/m^3 of water, they charge the buyers between INR 1.5/m^3 and INR 2/m^3 (US$1 equates to INR 50). Similar trends were found in the Kolar district, where well owners charge up to INR 6.5/m^3 (see Deepak et al., 2005), against a close to zero marginal cost of pumping groundwater. In many areas, groundwater-intensive use leads to water quality deterioration, causing scarcity of safe water for drinking. In such situations, the draft does not necessarily exceed the recharge. While the issue is of salinity in coastal Saurashtra and Chennai, it is arsenic content in deep aquifers in West Bengal (Kumar and Shah, 2004).

Groundwater over-use, like the use of other natural resources, involves ethical considerations (Custodio, 2000). These mainly revolve around the distribution of benefits and costs of water use and the risks associated with it (Llamas and Priscoli, 2000). The extent to which wasteful use practices are involved in major sectors of water use and the degree to which water abstraction practices reduce the opportunities of users – a neighbouring farmer, the individual himself, and others – are the major issues to be investigated (Kumar et al., 2001). In a water-scarce region, physically and economically inefficient uses should be discouraged. Contrary to this, even in regions where acute scarcity of groundwater exists, farmers use traditional irrigation methods that are wasteful, and allocate water to economically inefficient uses (see Deepak et al., 2005; Kumar, 2005). In hard rock areas, competitive drilling by powerful farmers causes a reduction in the yield of neighbouring wells due to well interference, depriving resource-poor farmers (Janakarajan, 2002; Deepak et al., 2005).

To sum up, the current assessment of groundwater over-exploitation does not give a clear picture of the actual intensity of over-exploitation in both absolute and relative terms. It tends to underestimate the magnitude of groundwater over-exploitation in India, which can be assessed from the negative social, economic and ecological consequences of over-development. From that perspective, many districts in Madhya Pradesh, Andhra Pradesh and Tamil Nadu could be actually over-exploited, though the official figures show that they fall under "safe", "semi-critical" or "critical" categories. The regions which have serious problems are alluvial Punjab, both the hard and alluvial areas of Gujarat, and the hard rock areas of Maharashtra, Tamil Nadu, Karnataka and Andhra Pradesh.

Impact of groundwater over-exploitation on agriculture and food security

The past decades have witnessed a slow down in growth of agricultural GDP from 3.3 per cent during 1980–95 to 2 per cent during 1995–03 (GOI, 2008). This has been accompanied by a decline in food consumption per unit of population. The per capita net availability of food grains (a rough measure of consumption) in 2004–06 was 7.8 per cent lower than in 1994–96. The agricultural crisis has grave implications for the country's ability to feed itself. In order to maintain the per capita production level of 2001–02, when it touched an all time high of 207.3 kg per person, food grain production should have touched 240 million tonnes in 2009–10. However, it was only 218.20 million tonnes, down from 227.9 million tonnes in 2008–09. The sharp decline was attributed to droughts in 2009, which impacts on the availability of water in reservoirs, soil profile and wells.

As groundwater contributes more than 5 per cent of India's GDP (Kumar, 2007) and accounted for nearly 61.2 per cent of the net irrigated area in the year 2000 (Ministry of Agriculture, Government of India), it is a truism that depletion will have a long-term impact on the country's economic growth and food security. But the potential future impact of groundwater over-exploitation in a particular region on India's food security depends on the relative contribution of well irrigation in that region to India's food security; the degree of over-exploitation of groundwater in the region; and the degree of vulnerability of the region. Here, vulnerability is considered to be an inverse function of the groundwater stock available for mining and the amount of water import available from outside that help improve the condition of groundwater. From that point of view, alluvial Punjab can be considered less vulnerable, though the degree of over-exploitation is very high as per our criteria.

According to some estimates, groundwater accounts for nearly 80 per cent of the agriculture production from irrigated areas in the country. Its contribution to the nation's food basket is quite major. However, the relative contribution of groundwater to the irrigated area varies widely from state to state. Table 3.4

Table 3.4 Gross irrigated area and well irrigated area for major Indian states

Sr. no.	*Name of the state*	*Gross irrigated area*	*Gross groundwater irrigated area*	*Percentage contribution of groundwater*
1	Andhra Pradesh	5.74	2.45	42.68
2	Gujarat	3.51	2.81	80.06
3	Haryana	5.22	2.57	49.23
4	Karnataka	3.17	1.19	37.54
5	Madhya Pradesh	4.59	3.10	67.54
6	Maharashtra	3.82	2.63	68.85
7	Odisha	2.39	0.62	25.94
8	Punjab	7.80	5.92	75.90
9	Rajasthan	6.60	4.30	65.15
10	Tamil Nadu	3.50	1.88	53.71
11	Uttar Pradesh	17.67	13.42	75.95
12	West Bengal	3.50	2.13	60.86

Source: Ministry of Agriculture, Government of India, 2000.

provides the aggregate area and percentage area irrigated by groundwater in major Indian states. It clearly shows that, in the semi-arid and arid states, the aggregate area under well irrigation is very large. Also, the percentage contribution of groundwater to total irrigation is major.

The problems of groundwater depletion are encountered in both alluvial areas and hard rock areas, including, for example, the alluvial areas of Punjab, Haryana and the Gujarat mainland, and the hard rock areas of Andhra Pradesh, Tamil Nadu, Karnataka and the Saurashtra region of Gujarat. These regions also show much higher rates of withdrawal of groundwater in per capita terms when compared to some of the physically water-rich regions (Figure 3.3). This sharp difference could be attributed to the differences in per capita cropped land, and climatic conditions, which change the demand for water for crop growth. Punjab has the highest rate of withdrawal of groundwater with a per capita annual draft of nearly 1279.2 m^3, and the north-eastern states have the lowest (15.9 m^3). The rich groundwater endowment in the extensive alluvium extending over most parts of Punjab support intensive cultivation of the land with high water-intensive crops such as wheat and paddy in this region notwithstanding its arid and semi-arid climatic conditions.

Within alluvial areas, over 80 per cent of the blocks in Punjab and Rajasthan, and over 60 per cent of the blocks in Haryana are falling in either over-exploited or critical or semi-critical blocks (GOI, 2005). With secular decline in water levels, shallow wells dry up. As the investment for drilling tube wells reaches astronomical heights, the poor farmers lose out in the race for water. They are either forced

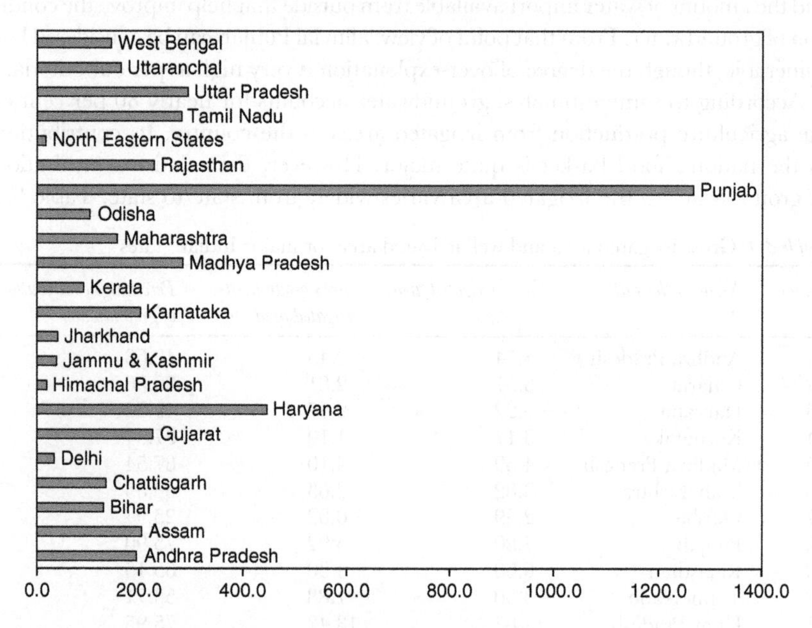

Figure 3.3 Per capita groundwater withdrawal rates in different states (m^3/annum).

Source: authors' own estimates based on GOI, 2001 and 2005.

to purchase water from the rich well owners at prohibitive prices, or shift to rain-fed farming practices. For instance, the tube well owners of Mehsana in north Gujarat charge as high as INR 70–INR 100 for an hour of irrigation service. This means that the economics of farming is adversely affected due to the rise in cost of production, which in turn affects livelihood security. The water buyers show an increasing tendency to grow cash crops that give much higher returns per unit of water consumed, as they are confronted with the high marginal cost of using water and have limited access to irrigation water in volumetric terms (Kumar, 2005).

It is important to note that the alluvial areas that fall under a semi-arid climate, such as Punjab, Haryana, Rajasthan and north Gujarat, are large exporters of agricultural commodities. While Punjab and Haryana exports cereals such as wheat and rice, Rajasthan exports wheat (Amarasinghe et al., 2004), and north Gujarat is a net exporter of milk (Singh et al., 2004). Table 3.5 shows the aggregate and relative contribution to India's wheat and rice production of five Indian states falling under alluvial areas. It can be seen that a little more than 44 per cent of the total wheat production in the country comes from the three north Indian states, which are also known for severe problems with groundwater over-draft. These states also contribute more than 20 per cent of India's rice production. Though their relative contribution is not high, the fact that they are largely wheat-consuming states means that most of their production is available for export to the rice-consuming states.

Therefore, permanent depletion of groundwater in these regions would adversely affect national food production with the area under well irrigated crops reducing and farmers moving away from cereals to less water-consuming and high risk cash crops. The situation is likely to be more severe in states such as Rajasthan and Gujarat, where replenishment of groundwater in the over-exploited areas through import of surface water is extremely limited.

Another probable consequence of depletion is that crops become highly vulnerable to the vagaries of the monsoon, with widespread failure during droughts. As noted by Kumar et al. (2006), high inter-annual variability in rainfall and frequent droughts are characteristic features of these low-medium rainfall regions. As a result, agriculture and the rural economy become more and more vulnerable to droughts. The rich farmers are able to sustain tube well irrigation because of highly subsidized electricity. They also pump out extra water and provide irrigation services to the neighbouring farmers and thereby earn income.

Table 3.5 Aggregate and relative contribution of states falling in semi-arid alluvial areas to India's wheat and rice production (2000)

Sr. no.	*Name of the state*	*Wheat production (million tonnes)*	*Percentage contribution*	*Rice production[1] (million tonnes)*	*Percentage contribution*
1	Punjab	16.01	22.10	15.207	10.79
2	Rajasthan	6.36	8.78		
3	Haryana	9.79	13.51	5.057	3.63
	Total of all Indian production	72.44		139.13	

Source: Ministry of Agriculture, Government of India.

Again, when groundwater resources deplete and the cost of well construction and pumping increases, the system of trading water provides greater economic opportunities to well owners having large holdings, and lesser opportunities to well owners having smaller holdings and to water buyers. This is due to the fact that, for a large farmer, the implicit unit cost of water is much lower as compared to small farmers. At the same time, a small farmer will not be able to raise the water charges to match the implicit cost of pumping, as the prices are determined by market forces (Kumar et al., 2001).

Analysis has shown that in the deep tube well areas of north Gujarat, if the state electricity boards start charging the full cost of electricity for pumping, the irrigated production of many crops would be unviable (IRMA/UNICEF, 2001; Kumar and Singh, 2001). This means that, from a larger societal point of view, groundwater irrigation in such situations does not contribute to economic growth. On the other hand, it also has negative ecological impact. The cumulative effect will be that the net social welfare is negative.

In hard rock areas, as seen earlier in this chapter, one of the immediate consequences of over-development has been the increase in the incidence of well failures, a reduction in well yields and shrinkage in their commands (Kumar, 2007). In such cases, farmers are found to drill bore wells in order to sustain access to irrigation water.

Here, as well, the poor farmers lose out in the race. This has led to a widespread emergence of monopolistic groundwater markets (Janakarajan, 2002; Deepak et al., 2005). Gradually, irrigated farming itself becomes unviable for water buyers. Wherever farmers continue irrigation with purchased water, shifting to high valued crops has been a widespread phenomenon. Hard rock areas contribute to India's food security in a major way. For instance, nearly 51.5 per cent of India's total rice production comes from the five states that fall into the hard rock category and which are facing the negative consequences of over-exploitation (Table 3.6). More importantly, Andhra Pradesh and Tamil Nadu, which are experiencing over-exploitation problems, account for 20.2 per cent of India's rice

Table 3.6 Contribution of states falling in hard rock areas to India's (rough) rice production in 2006

Sr. no.	*Name of the state*	*Rice production (million tons)*	*Percentage contribution to total production*	*Wheat production (million tons)*	*Percentage contribution*
1	Andhra Pradesh	17.796	12.79	0.00	0.00
2	Chattisgarh	7.562	3.51	0.00	0.00
3	Karnataka	4.893	5.43	0.00	0.00
4	Madhya Pradesh	2.052	2.72	4.86	6.70
5	Maharashtra	3.794	7.38	1.18	1.62
6	Odisha	10.191	7.32	0.00	0.00
7	Tamil Nadu	10.263	12.79	0.00	0.00
	All India total	139.13		72.44	

Source: rough rice production in India, by state, 1961–2006; data sourced from various reports by the Directorate of Economics and Statistics and the Ministry of Agriculture, Government of India.

production. Hence the impact on food security of depletion in hard rock areas would be significant.

In a nutshell, groundwater depletion and resultant water scarcity would have multiple effects. First, it would reduce the contribution to the nation's grain pool as the states experiencing depletion have significantly large well irrigated areas, and are exporters of grains and agricultural commodities. Decline in food production could increase domestic food prices. Further, depletion would force the farmers, particularly water buyers, to grow high-valued crops that are often risky. As the prices of these high valued crops are sensitive to market fluctuations, the farmer households also become vulnerable. This would adversely affect their domestic food security.

Opinions vs. realities

As discussed in Chapter 2 of this book, the fact that, at the aggregate level, only a little more than 50 per cent of the dynamic groundwater resources are exploited, has made at least some researchers argue for intensive well irrigation in eastern India, which is groundwater abundant, yet subject to agricultural stagnation (IWMI, 2007; Sharma, 2009). Furthermore, informal pump rental markets have been suggested as institutional mechanisms to promote access equity in groundwater irrigation (Shah, 2001; Mukherji, 2003; Mukherji et al., 2009).

Suggestions for flat rate pricing of electricity have also been made, in order to promote water markets and reduce the monopoly power of well owners (Shah, 2001; Mukherji et al., 2009). But in water-abundant eastern India, electric pump owners enjoy higher monopoly power over buyers in water trading (Kishore, 2004). The monopoly power comes from the poor transferability of water, and the very high transaction cost of obtaining power connections. Hence, power pricing policy will have very little impact on the price of water unless the issue of monopoly is addressed. The impact would also be nonexistent if a significant number of well owners in the locality continued to use diesel engines, incurring higher costs in the production of water. The real challenge lies in a) making the process of securing electricity connections in farm sector easy; and b) making access to pump subsidies easy for poor small and marginal farmers (Kumar, 2007). Even if these issues are addressed, the constraint imposed by land availability in enhancing production cannot be ignored. In fact, the intensity of cultivation and irrigation are already considerably high in eastern India (GOI, 1999), and there are ecological constraints in any further expansion of cultivated and irrigated areas.

Another alternative suggested by many researchers seeking to sustain the boom in well irrigation is water harvesting and the artificial recharging of groundwater to arrest depletion (Shah et al., 2003; Sharma, 2009). Research has shown that the potential for artificial recharge in India is very low in these arid and semi-arid regions facing over-draft. The reasons are a) low to medium mean annual rainfalls and highly variable and erratic rainfall, which reduces the overall runoff availability, runoff collection efficiency, and hence economic viability; b) the poor infiltration capacity of thin soils in the hard rocks areas that constitute a major proportion of India's semi-arid regions, resulting in poor recharge rates (Muralidharan and Athawale, 1998); c) the poor storage potential of hard rock aquifers underlying the areas; and d) high rates of evaporation from water bodies (Kumar et al., 2006; Kumar et al., 2008a). Analysis also shows that intensive water harvesting activities would have serious negative ecological consequences in arid and semi-arid regions (Kumar et al., 2006; Kumar et al., 2008a).

A virtual water trade has been suggested, by some scholars, as a means to deal with the groundwater crisis in water-scarce regions (Iyer, 2008; Sharma, 2009). Their argument iss that the ecologically fragile regions, such as Punjab, are producing water-intensive crops at the cost of resource use sustainability and energy use efficiency (Gulati, 2002), and that, instead, food security can be achieved by encouraging water-rich regions such as eastern India to produce surplus food for other regions (Sharma, 2009). The proponents of this virtual water trade argue that eastern India has a comparative advantage in producing wheat and rice with much less water. It goes without saying that the water-scarce regions need to make agricultural water use much more efficient. But the latter part of the argument is far fetched. It missed the point that the water-rich regions lack sufficient amount of arable land that can be put to use for producing sufficient food for themselves, and that only the water-scarce regions are endowed with a sufficient amount of arable land to produce a surplus of food. Analysis shows that the virtual water trade is governed by access to arable land, and not renewable water resources. While access to arable land ensures some amount of water in the soil profile to grow crops, "water richness" does not ensure the amount of land to put that water to use (Kumar and Singh, 2005).

Alternatives for sustaining groundwater irrigation for food security

In many semi-arid and arid parts of the country, groundwater irrigation has expanded considerably during the past few decades (Sharma, 2009). Most of these regions are underlain by hard rock formations with poor groundwater potential. They include parts of western and central Odisha, Karnataka, Tamil Nadu, Andhra Pradesh, Maharashtra, Chattisgarh and most parts of Madhya Pradesh. Alhough according to official statistics groundwater development is still very low in these regions (GOI, 2005), it is based on average annual abstraction and recharge estimates.

The outflows into rivers and streams, which reduce the utilizable groundwater from what is available through recharge, are quite remarkable in the upper catchments of many of those river basins that are underlain by hard rocks. As a result, official figures still identify many of these areas as suitable for further exploitation (Kumar, 2007). Due to acute groundwater scarcity, farmers in these areas have already moved from open dug wells to deep bore wells, and the opportunities for expanding the well irrigated area are limited. However, these factors are not taken into account in the estimation of groundwater irrigation potential by central and state agencies (Kumar and Singh, 2008).

In alluvial Punjab and Haryana, the net sown area, the net irrigated area and irrigation intensities are already at their peak, and further expansion is not possible.

In the recent past, groundwater irrigation has increased in sub-humid regions including eastern Uttar Pradesh, Bihar, Assam and West Bengal. There is plenty of both dynamic and static groundwater in this deep alluvial region (GOI, 2008). These regions can support more wells. The issue is of availability of a sufficient amount of arable land to expand the net sown area, as irrigation intensities are already very high. This is also the case with UP and West Bengal. UP has the largest area under well irrigation among all Indian states (Kumar, 2007). In Bihar, the potential for increasing the irrigated area is very low due to several ecological factors, though there could be a remarkable increase in surplus value product from agriculture through a reduction in the cost of irrigation, and improvements in the quality of irrigation, particularly for water buyers (Kishore, 2004).

Kumar (2007) discussed four major ways to sustain the groundwater economy.

1. Improve the allocation of surface irrigation in intensively irrigated areas facing over-exploitation.
2. Improve the efficiency of utilization of green water or the rainwater held in the soil profile (P_e).
3. Reduce the soil water depletion through a reduction in the amount of residual moisture held in soils after harvesting.
4. Reduce the consumptive use of water (ET) through a shift to low water consuming crops that are economically more efficient, i.e., crops that give higher net returns per every unit of water consumed (INR/ET).

A fifth way, identified by Sharma (2009) is to delay the sowing of crops to reduce the PET.

We now turn to address these one by one.

In some of the intensively irrigated semi-arid areas, the scope for increasing the allocation of surface water needs to be explored to improve the groundwater balance and sustain well irrigation. In other semi-arid areas, where the proportion of the area to the total cultivated areas are small, surface water allocation should be to improve groundwater balance and to raise the area under irrigation above the current levels. This is particularly important for boosting food production, as indicated by the analysis presented above, in that the yield impact of irrigation on food crops is generally high in these semi-arid tropics as compared to the subtropical regions.

Furthermore, an increase in the efficiency of the application of water, through better reliability of water delivery and control over the water application, could help reduce the amount of water depleted, and increase the consumptive use and productivity of water (Kumar and van Dam, 2009). Micro irrigation is one technology suitable for this (Narayanamoorthy, 2004; Kumar, 2009b, 2010). However, when there is plenty of surface water available for irrigation, particularly during the monsoon, the application of excessive irrigation and the use of the return flows for recharge can be explored.

Improving the efficiency of use of soil water offers tremendous potential in medium and high rainfall areas in reducing the irrigation requirement of many crops. This can be done through advancing the sowing of monsoon crops; conservation of available soil moisture through evaporation prevention; and in situ water harvesting. Shifting to orchard crops and oil seeds can bring down the evapo-transpirative requirement of crops. However, in regions like Punjab, where almost all the cultivated land is irrigated, it is important that such crops give the same or higher return per every unit of land irrigated, too. The reason is that the scope for expanding the area under the crop won't be available in these regions due to the present intensive cultivation.

But there are policy constraints to achieving water use efficiency improvements, namely economically inefficient ways of pricing canal water and power consumption in agriculture. Under the current modes of pricing followed by most Indian states, the marginal costs of using water and electricity are close to zero, leaving little incentive among farmers to adopt measures to improve the efficiency of water use in irrigation. Therefore it is important to introduce volumetric pricing of canal water, and pro rata pricing of electricity used in groundwater (Kumar, 2009a). But, as noted by Kumar and van Dam (2009), while volumetric pricing can definitely bring about efficiency improvements in water use, it may not affect any reduction in irrigation water demand in situations where farmers have the opportunity to expand the area under irrigation. Hence, rationing of the volume of water delivered to the field also might be required in such situations (Kumar and van Dam, 2009). Delayed sowing of rice, as tried in Punjab, can reduce the PET and increase the availability of soil moisture from rainfall, thereby reducing the irrigation water requirements (Sharma, 2009).

Conclusions

Food crisis is as much a crisis of land in water-rich regions, as it is a crisis of water in semi-arid and arid, water-scarce, regions. Groundwater over-draft problems in the water-scarce regions increase the magnitude of the crisis. In order to understand the real magnitude of groundwater over-exploitation problems in India, the assessment should involve complex considerations that are hydrological, hydrodynamic, economic, social and ethical in nature. Combining official statistics of groundwater development in the country with detailed information on groundwater balance, geology, water level fluctuations and the negative consequences of groundwater-intensive use, such as well failures, reduction in well yields and cost of groundwater abstraction, highlights a far more serious problem of resource over-exploitation than that indicated by official assessments.

If unchecked, its impact on national food security is likely to be severe, as the regions that are experiencing over-draft are also regions producing surplus cereals that are exported to land-starved water-surplus regions. The food security impact would be multiple. First, the groundwater depletion shrinks the area irrigated by wells, thereby decreasing the area irrigated under cereals. Second, when water becomes scarce, and the cost of irrigation water rises, the farmers move away from traditional cereal crops that give low returns per unit of water and adopt cash crops that are high risk. This also can lead to a decline in food production and national food security. All these could lead to rising prices of cereals, jeopardizing the ability of poor people to purchase food. As the prices of the high valued crops are highly sensitive to market fluctuations, the farmer households can also become vulnerable to income losses, thereby exposing them to food insecurity.

Several of the ideas being pursued by policy makers, or discussed as solutions to groundwater over-exploitation problems, are naive. They need to give way to serious scientific research on the physical and economic viability and the actual potential of the five strategies (discussed above) at the macro level. What is most important is to introduce reforms in the water and energy sector, including volumetric pricing of canal water, volumetric water allocation, and pro rata pricing of electricity used in groundwater (Kumar, 2009a, 2010).

Practical policy interventions

Two of the major policy interventions required to implement the strategies discussed above are in the water and electricity sector: to improve water use efficiency and energy use efficiency, in agriculture. This will need to be done at the level of the state governments concerned. Technical interventions are also required, to introduce tariff reforms in water and electricity, including devices for measurement of the volume of water delivered from canals to the farmers' fields and meters for the measurement of electricity consumption for groundwater pumping. Institutional reforms would be required to introduce volumetric rationing of canal water or water entitlements. Water pricing and energy pricing decisions are political, and have serious social and political ramifications. In the context of electricity pricing, Kumar (2009b) notes that such proposals get rejected on flimsy grounds (Kumar, 2009b). What is important is that the opportunity cost of not doing this is significant, in the form of low agricultural productivity and threats to the sustainability of groundwater resources and, thus, the livelihoods of millions of farm households. On the other hand, there are economic benefits to introducing pro rata pricing and charging for irrigation water volumetrically, if followed by better quality supply. These benefits include greater agricultural outputs and improved financial viability of both irrigation and the power sector, and higher income for farmers.

Investment in inter-basin water transfer projects, to enable transfer of water from (physically) water surplus basins to (physically) water-scarce basins, would enable the expansion of the irrigated area in water-scarce regions. This would also call for policy changes, but at the level of the federal government, as in most cases more than one state would be involved in such a project. There are major scientific, legal, social, environmental, ecological, financial and engineering issues involved in inter-basin water transfer projects (Kumar et al., 2008b). Improving the electricity supply infrastructure in eastern India, particularly Bihar, also requires public policy intervention (Kumar, 2007).

Note

1 Figures of raw rice production are for the year 2006.

References

Amarasinghe, U. A., Shah, T., & Singh, O. P. (2007). *Changing consumptions patterns: implications on food and water demand in India.* Research report 119. Colombo, Sri Lanka: International Water Management Institute.

Amarasinghe, U. A., Sharma, B. R., Aloysius, N., Scott, C. A., Smakhtin, V., de Fraiture, C., Sinha, A. K., & Shukla, A. K. (2004). *Spatial variation in water supply and demand across river basins of India.* Research Report 83. Colombo, Sri Lanka: International Water Management Institute.

Chowdhury, A. (2011). Food price hikes: How much is due to excessive speculation? *Economic and Political Weekly, 46* (28), 12–15.

Custodio, E. (2000). *The complex concept of groundwater exploitation.* Papeles del Proyecto Aguas Subterráneas A1. Santander, Spain: Fundacion Marcelino Botin.

Deepak, S. C., Chandrakanth, M. G., & Nagaraj, N. (2005). *Water demand management: A strategy to deal with water scarcity.* Paper presented at the 4th annual partners' meet of the IWMI-Tata Water Policy Research Program, Anand, Gujarat, India, February.

Evenson, R. E., Pray, C. E., & Rosegrant, M. W. (1999). Agricultural research and productivity growth in India. *Research Report 109.* Washington, DC, USA: International Food Policy Research Institute.

Falkenmark, M. (2004). Towards integrated catchment management: Opening the paradigm locks between hydrology, ecology and policy making. *International Journal of Water Resources Development, 20* (3), 275–282.

Government of India. (1999). *Integrated water resource development: A plan for action. Report of the National Commission on Integrated Water Resources Development, vol. I.* New Delhi, India: Ministry of Water Resources, Government of India.

Government of India. (2000). *Agricultural census 2000.* New Delhi, India: Government of India.

Government of India. (2001). *Population census 2001.* New Delhi, India: Government of India.

Government of India. (2005). *Dynamic ground water resources of India.* New Delhi, India: Central Ground Water Board, Ministry of Water Resources, Government of India.

Government of India. (2008). *11th five year plan, 2007–2012: Agriculture, rural development, industry, services and infrastructure.* New Delhi, India: Planning Commission, Government of India.

Gulati, A. (2002). *Challenges to Punjab agriculture in a globalizing world.* Paper based on the presentation given at the policy dialogue on challenges to Punjab agriculture in a globalizing world, IFPRI and ICRIER, New Delhi, India.

Institute of Rural Management Anand & UNICEF. (2001). *White paper on water in Gujarat.* Report submitted to the Government of Gujarat. Anand, Gujarat, India: Institute of Rural Management.

International Water Management Institute. (2007). *Water figures: Turning research into development.* Newsletter 3. Colombo, Sri Lanka: International Water Management Institute.

Iyer, R. R. (2008). Water: A critique of three basic concepts. *Economic and Political Weekly, 43* (1), 15–18.

Janakarajan, S. (2002). *Wells and illfare: An overview of groundwater use and abuse in Tamil Nadu, south India*. Paper presented at the 1st annual partners' meet of the IWMI-Tata Water Policy Research Program, Anand, Gujarat, India.

Kishore, A. (2004). Understanding agrarian impasse in Bihar. *Economic and Political Weekly*, *39* (31), 3484–3491.

Kumar, M. D. (2003). *Food security and sustainable agriculture in India: The water management challenge*. Working paper 60. Colombo, Sri Lanka: International Water Management Institute.

Kumar, M. D. (2005). Impact of electricity prices and volumetric water allocation on energy and groundwater demand management: Analysis from western India. *Energy Policy*, 33 (1), 39–51.

Kumar, M. D. (2007). *Groundwater management in India: Physical, institutional and policy alternatives*. New Delhi, India: Sage Publications.

Kumar, M. D. (2009a) *Water management in India: What works, what doesn't*. New Delhi, India: Gyan Publishing.

Kumar, M. D. (2009b). Opportunities and constraints to improving water productivity in India. In M. D. Kumar & U. A. Amarasinghe (Eds.), *Water productivity improvements in Indian agriculture: Potentials, constraints and prospects*. Strategic analysis of national river linking project (NRLP) of India series 4. Colombo, Sri Lanka: International Water Management Institute.

Kumar, M. D. (2010). *Managing water in river basins: Hydrology, economics and institutions*. New Delhi, India: Oxford University Press.

Kumar, M. D., & Shah, T. (2004). Groundwater pollution and contamination in India: The emerging challenge. *Hindu Survey of Environment 2004*, 7–12.

Kumar, M. D., & Singh O. P. (2001). *Rationed supplies and irrational use: Analysing water accounts of Sabarmati basin*. Monograph 1. Anand, Gujarat, India: India Natural Resource Economics and Management Foundation.

Kumar, M. D., & Singh, O. P. (2005). Virtual water in global food and water policy making: Is there a need for rethinking? *Water Resources Management*, *19* (6), 759–789.

Kumar, M. D., & Singh, O. P. (2008). How serious are groundwater over-exploitation problems in India: Fresh investigations into an old issue. Paper presented at the 7th annual partners' meet of the IWMI-Tata Water Policy Research Program, Anand, Gujarat, India.

Kumar, M. D., & van Dam, J. (2009). Improving water productivity in agriculture in India: Beyond more crop per drop. In M. D. Kumar & U. A. Amarasinghe (Eds.), *Water productivity improvements in Indian agriculture: Potentials, constraints and prospects*. Strategic analysis of national river linking project (NRLP) of India series 4. Colombo, Sri Lanka: International Water Management Institute.

Kumar, M. D., Ghosh, S., Patel, A., Singh, O. P., & Ravindranath, R. (2006). Rainwater harvesting in India: Some critical issue for basin planning and research. *Land Use and Water Resources Research*, *6* (1), 1–17.

Kumar, M. D., Malla, A. K., & Tripathy, S. (2008a). Economic value of water in agriculture: Comparative analysis of a water-scarce and a water-rich region in India. *Water International*, *33* (2) 214–230.

Kumar, M. D., Patel, A., Ravindranath, R., & Singh, O. P. (2008b). Chasing a mirage: Water harvesting and artificial recharge in naturally water-scarce regions. *Economic and Political Weekly*, *43* (35), 61–71.

Kumar, M. D., Singh, O. P., & Singh, K. (2001). *Groundwater depletion and its socioeconomic and ecological consequences in Sabarmati river basin*. Monograph 2. Anand, Gujarat, India: India Natural Resource Economics and Management Foundation.

Llamas, M. R., & Priscoli, J. D. (2000). *Report of the UNESCO working group on ethics of freshwater use.* Papeles del Proyecto Aguas Subterráneas. Santander, Spain: Fundacion Marcelino Botin.

MS Swaminathan Research Foundation & World Food Programme (2010). *Report on the status of food insecurity in urban India.* Chennai, India: MSSRF.

Mukherji, A. (2003). Groundwater development and agrarian change in Eastern India, *IWMI-Tata Comment, 9,* 11.

Mukherji, A., Villholth, K. G., Sharma, B. R., & Wang, J. (2009). *Groundwater governannce in the Indo-Gangetic and Yellow River basins: Realities and challenges.* Boca Raton, USA: CRC Press (Taylor & Francis Group).

Muralidharan, D., & Athawale, R. N. (1998). *Artificial recharge in India.* Base paper prepared for Rajiv Gandhi National Drinking Water Mission, Ministry of Rural Areas and Development, National Geophysical Research Institute, Hyderabad, India.

Narayanamoorthy, A. (2004). Drip irrigation in India: Can it solve water scarcity? *Water Policy, 6,* 117–130.

National Rain-fed Area Authority. (2011). *Challenges of food security and its management.* New Delhi, India: National Rain-fed Area Authority.

Pisharoty, P. R. (1990). *Characteristics of Indian rainfall.* Monograph. Ahmedabad, India: Physical Research Laboratories.

Planning Commission. (2007). *Report of the steering committee on agriculture and allied sectors for the preparation of 11th five year plan (2007–2012).* New Delhi, India: Planning Commission, Government of India.

Shah, T. (2001). *Wells and welfare in Ganga basin: Public policy and private initiative in eastern Uttar Pradesh, India.* Research report 54. Colombo, Sri Lanka: International Water Management Institute.

Shah, T., Debroy, A., Qureshi, A., & Wang, Z. (2003). Sustaining Asia's groundwater boom: An overview of issues and evidences. *Natural Resources Forum, 27* (2), 130–141.

Sharma, K. D. (2009). Groundwater management for food security in India. *Current Science, 96* (11), 1444–1447.

Singh, O. P., & Kumar, M. D. (2009). Impact of dairy farming on agricultural water productivity and irrigation water use. In M. D. Kumar & U. A. Amarasinghe (Eds.), *Water productivity improvements in Indian agriculture: Potentials, constraints and prospects.* Strategic analysis of national river linking project (NRLP) of India series 4. Colombo, Sri Lanka: International Water Management Institute.

Singh, O. P., Sharma, A., Singh, R., & Shah, T. (2004). Virtual water trade in the dairy economy: Analyses of irrigation water productivity in dairy production in Gujarat, India. *Economic and Political Weekly, 39* (31), 3492–3497.

Todd, D. K. (2003). *Groundwater hydrology* (2nd edn.). Singapore: John Wiley and Sons (Asia) Pte Ltd.

4 Redefining the objectives and criteria for the evaluation of large storages in developing economies

Zankhana Shah

Introduction

The present crisis in meeting the food, energy and water requirements of the world population has accentuated the 'dams or no dams' debate. Dams are being opposed on environmental (D'Souza, 2002; McCully, 1996), financial, economic and human rights fronts (Dharmadhikary, 2005), whereas the proponents of large dam push their agenda on the grounds of enhanced food and drinking water security, hydropower generation and flood control (Verghese, 2001; Vyas, 2001).

According to the World Register of Large Dams (2003) prepared by the International Commission on Large Dams (ICOLD), there are more than 47,000 large dams constructed across the world and another 1,700 dams under construction. These statistics are based on a definition that has dam height as the sole criterion. According to a World Commission of Dams database (2000), China has the highest number of large dams followed by the rest of Asia and North and Central America. But global comparisons provided by the ICOLD World Register of Large Dams (1998) based on storage volume show that nearly 29 per cent of the total storage from large dams (6,464 km^3) is in North America, followed by South America (16 per cent). China (10 per cent) stands fourth in this category. A lack of comprehensive criteria for defining "large dams" makes such statistics misleading.

There are inherent limitations in the methods used for benefit–cost (BC) analysis. The methods identify only those costs and benefits which can be assigned a market value. Many social and environment costs (as well as benefits) are not considered due to limitations in assigning them economic value. For example, a water resource planning exercise done in the Indian state of Gujarat has recommended the use of water from Sardar Sarovar Project (SSP) to recharge the regional phreatic aquifers (GOG, 1996 as cited in Ranade and Kumar, 2004). However, this was never considered in the original planning exercise.

The basic premise

The widely used criteria for defining large dams are not true reflections of the socio-economic and environmental concerns prevailing in developing economies, and therefore are not relevant. Definitions based on such poor criteria often invite strong reactions from the environmental lobby worldwide. Similarly, while there is a lot of advancement in the recent past in the BC calculations of dam projects, the methodologies still fail to capture the social and environmental benefits that are likely to accrue in future. Such benefits include drinking water security, groundwater recharge and a reduced cost of the energy required for pumping. Often, dam builders inflate certain components of the benefits and under-estimate some others, in order to pass through the scrutiny of national and international agencies. In the process, little attention is paid to alternative ways of designing dams.

The objectives of this chapter are a) to discuss the criteria used by various national and international agencies in defining large dams, and identify their limitations in the context of developing countries; b) to evolve meaningful criteria for defining large storages, which adequately integrate the growing social and environmental concerns associated with dam building; and c) identify the gaps in the current BC calculations, and set out new objectives and criteria for evaluating the impact of large dams in developing economies.

Large dams: history, definitions and recent trends

Definitions of large dams

Numerous definitions are available of large dams, each serving different purposes and objectives and, therefore, based on different criteria. According to the US Fish and Wildlife Service's Dam Safety Program, small dams are structures of less than 40 feet in height or that impound less than 1,000 acre-feet of water; intermediate dams are structures of 40 to 100 feet in height or that impound 1,000 to 50,000 acre-feet of water; and large dams are structures of more than 100 feet in height or that impound more than 50,000 acre-feet of water (www.fws.gov).

The Central Water Commission (CWC), India, in its guidelines for safety inspections, has provided various definitions of dams on the basis of their size, gross storage and hydraulic head (CWC, 1987). Against this, the Planning Commission of India has defined a large irrigation project as the designed for irrigating more than 10,000 hectares (ha) of land.

The most widely accepted definition of large dams is given by ICOLD. The ICOLD defines a large dam as one having a wall height of more than 15 metres from the lowest general foundation to the crest. However, even dams between 10 and 15 metres in height could be classified as large dams if they satisfy at least any one of the following criteria (Rangachari et al., 2000). First, the crest length is more than 500 metres. Second, the reservoir capacity is more than one MCM. Third, the maximum flood discharge is more than 2000 m^3 per second. Fourth, the dam has complicated foundation problems. Fifth, an unusual design. Since this definition has been widely accepted, all world dams are usually evaluated on the basis of this definition.

A brief history of dam construction, ideologies and investments on dams in India

Table 4.1 provides statistics on large dams in India, based on ICOLD data.

Table 4.1 Large dams in India

Period	*Number of large dams*		
	Height of 15 m or greater	*10 to 14 m high including dams the height of which is not known*	*Total*
Up to 1900	32	13	45
1901–1947	135	127	262
1948–1970	489	254	743
1971–1990	1,564	1,066	2,630
1991–2001	265	82	347
Data not available	434	174	608
Total	2,919	1,716	4,635

Source: data derived from the World Register of Dams, 2003, ICOLD.

These figures, categorized on the basis of dam height, can be highly misleading. For instance, the first 2,919 dams create a storage space of 296.29 BCM, with a mean storage space of 101.5 MCM. However, the remaining 1,716 dams together create a storage space of only 6.29 BCM, with a mean storage space of 3.65 MCM. Thus, these are not really large dams from perspective of storage capacity.

The total storage created by all large dams in India is only 302.58 BCM with a mean storage capacity of 64.28 MCM. This, however, does not mean that these dams actually store and provide the stated quantity of water: many large dams in India do not get sufficient storage due to inadequate inflows from their catchments, and the figures of storage capacity are of gross storage, not live storage. The current total live storage capacity of reservoirs in India is only 214 BCM. It is increasingly reduced, for many reservoirs, because of silting (Thakkar and Bhattacharyya, 2006, based on State Reservoir Survey data).[1]

Compared to India, the US has 16,383 dams listed in its National Dams Register, including dams of less than 10 metres in height. Of these, only 1,735 dams are more than 15 metres high, and create a total storage of 140. 14 BCM with a mean storage of 80.8 MCM. The remaining 14,648 dams provide a total storage of 342 BCM with a mean storage of 23.3 MCM (estimates based on the US National Dams Register). This means that dams with height of less than 15 metres are very important storage systems for the US, as their total storage volume exceeds large dams.

In Australia, the mean storage of a large dam is 176.7 MCM (estimates based on data provided by the Natural Heritage Trust, 2000). In nutshell, though India appears to have more large dams, the water storage potential created by them is significantly lower than in many other countries.

Analysis of the defining criteria for large dams

The 4,635 large dams in India are either of a height above 15 metres or comply with any other criteria set out by the ICOLD definition. With the current technical excellence in the field of civil engineering and structural design, constructing a dam 15 metres or more high, or one with unusual design or a difficult foundation, is not challenging any more. Besides, criteria such as unusual design or a difficult foundation have not much to contribute towards environmental problems or water supply targets.

Should the sheer number of large dams really send warning signals on the magnitude of the costs being paid by society? To answer this question, it is crucial to know the relevance of the criteria used for classifying dams as "large". Most of the criteria for classifying dams as large or small evolved at times when the building of large dams used to pose major engineering challenges to humanity.[2] These criteria never tried to capture the social and environmental imperatives of building dams.

None of the definitions mentioned above, including that of ICOLD, is universally applicable. The various physical attributes of a dam such as height, storage volume and submergence area have different implications which are subject to change in accordance with dam location. Therefore, when we analyse the impact of dam construction, we cannot draw generalizations in respect of any of these attributes. For example, analysis of a dam on the basis its height does not always determine environmental impact, displacement or total storage volume and submergence area.

Normally, dam designers use the storage-elevation-area curve to determine the appropriate height of the dam and spillway capacity. Depending on the topography of the location, the storage-elevation-area relationships would change. In a deep gorge, the area under submergence of a high dam having a large storage volume may be very low. For example, the Idukki dam, which is a double curvature arch dam, is located in a deep gorge in Idukki in Kerala, India. It has a height of 555 feet and a storage volume of 2,000 MCM, and thus may not have submerged much area. An analysis of the data of 9,878 dams from ICOLD's World Register of Dams shows that the storage volume of a dam is not a function of its height (Figure 4.1).

Further analysis of ICOLD data shows that the area of land submerged by the reservoir is not a function of dam height (Figure 4.2).

Of the total 4,635 large dams of India – those with a height of more than 15 m or a storage volume higher than 1 MCM – 2,431 (more than 50 per cent) are built on local streams. Some of them might be tank systems with large surface areas. It is also possible that they are constructed under various small scale irrigation schemes

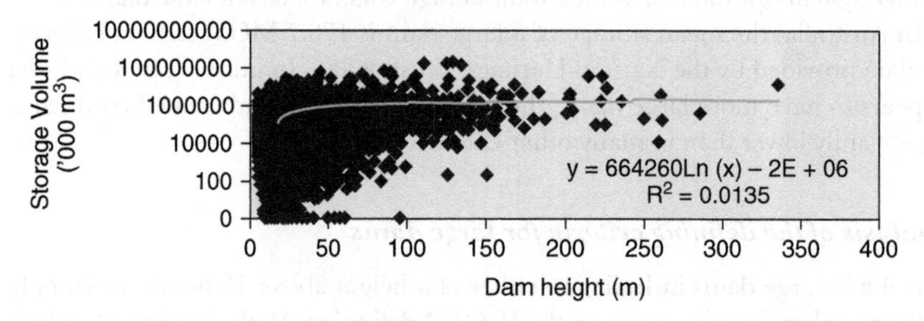

Figure 4.1 Dam height vs. storage volume.

Source: author's own analysis of ICOLD's data.

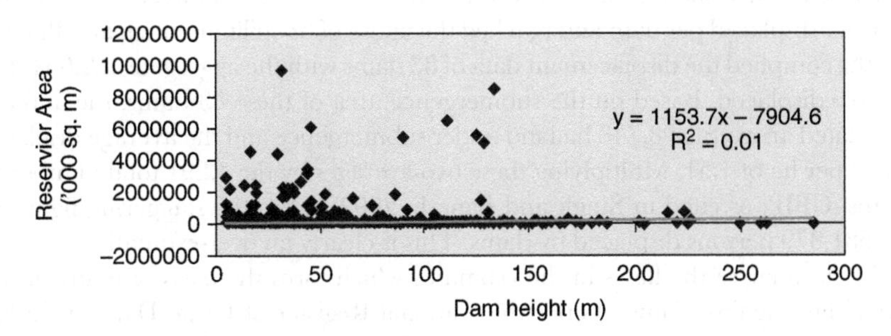

Figure 4.2 Dam height vs. reservoir area.

Source: author's own analysis of ICOLD data.

to achieve local benefits. Locally initiated water harvesting moves or small scale irrigation schemes usually do not cause displacement, and their negative social impact is also minimal to nil. That being the case, more than 50 per cent of India's large dams are socially and economically rewarding with minimum environmental costs. In fact, their presence might have contributed towards growth of vegetation, fisheries and water security.

Objectives and criteria for assessing large dams

Objectives and criteria for classifying large dams

It is evident from the above discussion that dam height does not have any bearing either on the area that dams submergeor the storage they create. On the other hand, the area submerged by dams has environmental, hydrological and socio-economic implications. It is also a good indicator of the potential ecological damage, though the actual consequences would depend on several factors, such as the nature of the eco-region and the population density where the dam is located.

Identifying the right kind of criterion for determining the impact of large dams also helps an assessment of the magnitude of the displacement issues. Global estimates of the magnitude of impact show that 40 to 80 million people have been displaced by dams (Bird and Wallace, 2001). In India, no authentic figures are available for dam-induced displacement. All available figures are estimates based on rough calculations. Fernandes et al. (1989) claimed that India had 21 million people displaced by dams. Some years ago, the (then) Secretary to the Ministry of Rural Development, Government of India, unofficially stated that the total number of persons displaced by development projects in India was around 50 million, and around 40 million of them were displaced by dams.

Some other estimates are based on average displacement per dam. Based on a study of 54 dams, the Indian Institute of Public Administration (IIPA) concluded that the average number of people displaced per dam was 44,182. Roy (1999) multiplied this figure by the 3,300 dams in India to yield a total of 145 million displaced. Since she felt this figure was too large, she took an average of 10,000 persons displaced per dam and reached the figure of 33 million. Singh and Banerji (2002) compiled the displacement data of 83 dams with the aggregate of 2,054,251 people displaced. Based on the submergence area of these 83 dams, the authors estimated an average 8,748 ha land under submergence and the average displacement per ha of 1.51. Multiplying these two averages by the 4,291 total number of dams (CBIP, as cited in Singh and Banerji, 2002) led to the staggering figure of 56,681,879 persons displaced by dams. This is clearly an over-estimation.

Let us analyse the flaws in the estimates which form the basis of many of the arguments against dams. As per the National Register of Large Dams in India, there are 1,529 large dams in the state of Maharashtra (CWC, 1994). As per the ICOLD criteria, the state has 1,700 large dams. If we adopt Roy's estimates of 10,000 persons displaced per large dam, Maharashtra alone should have displaced between 15.3 and 17 million people. It is most unlikely that such a large population in one state has no visibility. Given India's poor track record of rehabilitation, the majority of these displaced people might be facing poverty and impoverishment. On the contrary: Maharashtra stands next to Kerala on human development indicators (UNDP, 2006).

The major limitation of these estimates is that they are derived from an average displacement figure estimated per dam. The figures offered by CWC (1994), CBIP (1987, 1998) and the ICOLD World Register of Large Dams (2003) were on the basis of the total number of large dams per the ICOLD criterion. Thus, all the estimates of displacement have the inbuilt assumption that dam height influences displacement.

In practice, intensity of displacement would be largely determined by the submergence area and the population density of the region under consideration. The following figure supports the argument that it is a good indicator. It is based on an analysis of 156 large dams in India. It shows that the number of people displaced by dams increases linearly with an increase in submergence area. Submergence area explains displacement to the tune of 58 per cent (Figure 4.3). The rest could be explained by variation in population density. This is a high level of correlation, and therefore can be used to project the number of people displaced by dams.

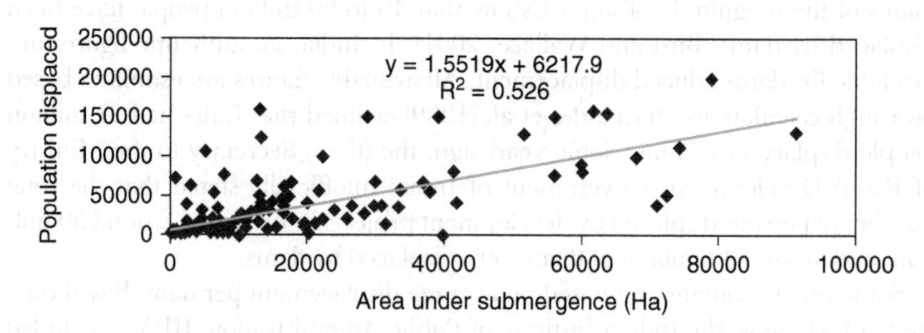

Figure 4.3 Submergence area vs. population displaced.

Source: Author's own analysis.

Figure 4.3 clearly shows that 100 ha of submerged land can cause the displacement of 150+ people. The total area submerged by 2,933 large dams in India (obtained from The World Register of Dams, 2003) was estimated to be 32,219.25 km². The area submerged by 4,635 dams was extrapolated to be 50,915 km². (32,219×4,635/2,933). Based on this, the total number of people displaced by dams was estimated to be 7.845 million people. This is far less than the displacement figures referred to earlier. The main utility of this relationship is that, once established for a given range of submergence area and population density, the number of people likely to be affected by dams could be estimated with a reasonable degree of accuracy.

Based on the above analysis, it can be inferred that a combination of parameters such as height, storage volume and submergence area would give a true reflection of the engineering, social and environmental challenges posed by a dam. Hence, criteria for classifying large dams should be developed through a consideration of all three important parameters.

New criteria for evaluating the performance of large dams

The economic impact of large dams in India is assumed to be negative, on the basis of construction cost over runs; poor performance of irrigation systems with heavy wastages due to poor conveyance efficiencies in the distribution system; negative downstream ecological impacts; preference for water-intensive and low water-efficient crops; water logging and salinity in command areas; and the problems of the over-estimation of benefits resulting from a shrinking of command areas due to the non-availability of water (Rangachari et al., 2000). In reality, very few studies comprehensively evaluate the long-term economic and social impact of large dams.

The criteria selected for impact evaluation also suffer from limitations, as usually they are no different to those used for BC calculations during the planning phase (Biswas and Tortajada, 2001). In the process, most evaluations under reported major benefits, such as food security from stable food prices, an increased rate of employment in agriculture, improved fisheries, increased access to drinking water supplies, development and growth of processing and marketing units, or improved groundwater balance due to return flows from canal irrigation (Hira and Khera, 2000). For instance, a recently concluded study on the Sardar Sarovar Project (SSP) covering seven districts of Gujarat estimated that the economic value of the indirect impact of canal irrigation on well irrigation through improved yield of wells ranged from a low of INR 24,719/ha in Panchmahals district to INR 113,587/ha in Ahmedabad district. Furthermore, for every hectare of well irrigated area in the SSP command, the indirect benefit to the society through energy saving in groundwater pumping ranged from a low of INR 768 in Bharuch to INR 9170 in Mehsana. This is in addition to the direct economic benefit from gravity irrigation, which was estimated to vary from a low of INR 24,903 per ha per annum in Panchmahals to INR 48,348 per ha in Bharuchper ha of gross cropped area (IRAP, 2012).

This is not to argue that large dam projects are free from problems. Many dams, especially those built in semi-arid and arid regions in India, are over-allocating water from their respective basins (Kumar, 2002). The irrigation agency was often keen to build oversized dams, taking the flows of low dependability as the design yield to inflate the design command and economic benefits. This leads to reduced flows or no flows in the downstream river, causing ecological problems (Kumar et al., 2000). But such problems have occurred primarily as a result of mismanagement and poor planning and cannot be labelled as a structural limitation of dams.

Criteria for evaluating environmental impact

The arguments against large dams often miss a scientific base. For instance, the long-held position of Narmada Bachao Andolan was that the social costs of large water development projects cannot be compensated by the increased economic benefits accrued from the use of water. They argued that complete rehabilitation of the displaced communities is impossible (Fisher, 2001), and that cheaper and easier options to large dams exist.

Such arguments gained credibility after the concept of virtual water trade was introduced in the early 1990s, with small water harvesting options gaining wide acceptance from then on. Some environmental activists advocate virtual water trade as an alternative to large dams, and suggest that water-scarce countries should import food grain from water-rich countries. In fact, many water-scarce regions in India export agricultural produce worth thousands of million cubic metres of water to regions that are water-rich (Amarasinghe et al., 2004; Singh, 2004). In a similar manner, local water harvesting solutions are found to be extremely limited in scope (Kumar et al., 2006).

Another criticism against large reservoir projects is water-logging and salinity problems in the command area. Part of the reason is that, world-wide, nearly 50 per cent of reservoir projects are for the purpose of irrigation. This has been an issue in many canal command areas of northern and north-western India and the Pakistan Punjab. Butdramatic changes in agriculture practices in these countries during the past couple of decades have converted some of these challenges into opportunities. With increasing groundwater draft for agriculture – which happened as a result of advancement in pumping technologies, massive rural electrification, and subsidized electricity for well irrigation – water-logging is becoming a non-issue. In Indian Punjab, the area under water-logging and salinity has actually reduced (Hira and Khera, 2000). The return flows from canals had played a significant role in sustaining tube well irrigation, and thereby agriculture, during the years of scarcity (Dhawan, 1990). An analysis by Kumar (2007) showed that nearly 5 per cent of the deep tube wells, 10 per cent of the dug wells and 5 per cent of the shallow tube wells in India are located in canal command areas. Induced recharge from canals controls groundwater mining for irrigation in many arid and semi-arid areas of India, thereby preventing the incidence of well failure.

While water-logging, salinity and downstream ecological damage are highly apparent when they occur, the unintended positive effects, such as drought proofing, drinking water security in rural and urban areas, increased biomass availability and increased inland culture fisheries, have been less talked about, and often less attributed to the dam construction. Their performance is not evaluated in relation to the number of jobs dams create in rural areas or the number of people who have benefited from drinking water availability.

Another unforeseen benefit is as follows: almost all major dams in the world are constructed for hydropower (Altinbilek, 2002), which is one of the cleanest powers in the world. Globally, nearly 19 per cent of all electric power is generated is from hydropower. Ideally, negative externalities created by thermal power on the environment could be treated as a positive externality of hydropower generation, so a kilowatt hour of energy produced from a hydropower plant should give an additional benefit equal to the cost of environmental damage caused by a thermal power plant generating same amount of power. In the case of SSP, the indirect economic impact of producing clean energy to the extent of 3,436 million units was estimated to be INR 1.61 billion per annum. This is the cost that will have to be incurred to mitigate the carbon emission that is likely to be caused by producing the same amount of energy through coal-based thermal power, the only alternative energy production system for the region (IRAP, 2012).

Criteria for evaluating ecological impact

The arguments about the downstream ecological impact of dams concern potential reduction in lean season flows (Smakhtin et al., 2004). In practice, regulatory reservoirs could be used to mimic th natural flows needed for ecosystem health. For instance, in the Narmada river basin in central India, large stretches between the Indira Sagar dam and the Sardar Sarovar dam will have higher lean season flows than would naturally be the case, due to flow regulations.

The more immediate, positive, ecological impact would be seen in water-starved regions when surplus flows from reservoir can be diverted for ecological uses. In the Narmada basin, the Sardar Sarovar reservoir is the terminal reservoir, and it can receive all surplus flows from the reservoirs upstream. This water is being diverted through the Narmada Main Canal which intersects rivers in north and central Gujarat, such as Sabarmati, Watrak, Shedhi, Meshwo, Khari, Rupen, Sipu and Banas (Desai and Joshi, 2008). Those stretches of the rivers which the canal intersects do not carry any flows during lean season. They now receive the excess flows diverted from the Sardar Sarovar reservoir, which provide for river ecosystems and also recharge the over-exploited alluvial aquifers of north Gujarat. In fact, access to water from the Sardar Sarover reservoir has played a significant role in increasing agricultural productivity of some arid and water-scarce regions of Gujarat as well as in achieving an impressive sectoral growth rate of 9.6 per cent at the state level (Kumar et al., 2010).

Criteria for evaluating social and economic benefits

According to Perry (2001), criteria such as food availability, food security, food prices and resettlement success are the right indicators to measure the economic performance of dams. Food security is an important water management goal for many water-scarce countries, including India and China (Kumar, 2003; Kumar and Singh, 2005). As per ICOLD data, nearly 48 per cent of all large dams worldwidewere built for irrigation. Still, the positive externalities induced by improved food security have been less articulated.

According to Bhalla and Mookerjee (2001), the total irrigation expenditure on major and medium irrigation schemes since independence in India has totalled INR 1, 87,000 crore (US$1 equates to INR 50) at 1999 prices. Against this, the total agricultural output in 1998–99 was close to INR 5,00,000 crore. Depending on the assumptions one makes for how much of the total investment is allocated for big dams (whether 100 per cent or 75 per cent) and depreciation rates (3 per cent to 5 per cent), one obtains IRRs in the range of 3 per cent to 9 per cent (Bhalla and Mookerjee, 2001). This is a direct benefit.

The positive externalities of increased food security should be assessed in terms of the opportunity cost of importing food. In order to examine how China and India influence international food prices, and in particular rising cereal imports due to increasing meat consumption (which is a response to income rises and declining domestic production due to degradation of the natural resource base), an IFPRI study used IMPACT (the International Model for Policy Analysis of Agricultural Commodities and Trade) to simulate a scenario of increased food imports by India of 24 million tonnes and China of 41 million tonnes in 2020. The model showed an increase in international wheat and maize prices of 9 per cent and rice prices of 26 per cent (Rosegrant et al., 2001).

If we consider the fact that half of the additional food grain of 94 million tonnes produced from irrigation in India since the 1950s is from large dams (Perry, 2001), and decide to compensate the reduced production in the absence of large dams through food imports, and we assume that prices increase from the present price by just US$20 per ton (US$1 equates INR 50), this would mean a total additional burden of INR 42.30 billion annually for the imported portion alone. This is more than 0.l0 per cent of India's GDP. If we assume that the current domestic cereal prices are close to import prices, the lower price consumers pay (say by US$20 per tonne) is the impact of domestic production of cereals on the food prices, and therefore can be considered as a positive externality of large dams.

The role oflarge dams in ensuring drinking water supplies in water-scarce regions is less appreciated. SSP in Western India, for example, is making a major dent in the rural and urban drinking water needs of 9,663 villages and 137 urban centres. Without the project, the drinking water situation in these areas would have been precarious due to an absence of any substitute source (Talati and Kumar, 2005). As many cities and towns are facing permanent depletion of local groundwater, many dams originally meant for irrigation are now supplying water for domestic consumption. While advocates of local alternatives for managing drinking water supplies had fiercely opposed regional water transfer from Narmada to Saurashtra and Kachchh in Gujarat on cost grounds, they have failed to demonstrate alternatives which are effective in both physically and economically (Kumar, 2004). In addition to the social impact of the provision of reliable supplies of water, in the form of reduction in distance travelled and time spent in water collection, there are significant indirect effects of the Narmada canal-based piped water supply. The indirect economic impact for a total population of 24.33 million people in Saurashtra, Kachchh and north and central Gujarat, owing to the saving in energy used for pumping groundwater for rural and urban water supply, was estimated to be INR 0.85 billion per annum (IRAP, 2012).

Access to water has a direct impact on human development (UNDP, 2006), and the role of large dams in increasing water security and reducing drought vulnerability is often undervalued in any assessment of their impact. A recent work involving analysis of data from 145 countries showed that improving the water security of a region, expressed in terms of the sustainable water use index (SWI), improves the human development by reducing mortality and malnutrition (Kumar and Mudgerikar, 2009). The analysis has also established that progress in human development has very little to do with the economic growth, and that it is possible to achieve good indicators of development even at low levels of economic growth, through investment in water infrastructure and welfare-oriented policies.

Internalizing the negative externalities in project costs

Water and power development agencies of poor countries in Asia and Africa show an unwillingness to include the negative externalities as part of the project cost. This is because they do not like to transfer those costs to the water users, due to fear that it would bring down the demand for water and thereby make the BC ratios very unattractive. Instead, the practice is to bundle all such costs, and come out with a compensation package for the affected people once the project passes scrutiny, by the donors, for economic viability.

This myopic tendency can be explained by the fact that the reduction in benefit resulting from the decision to cut down the size of the project to minimize the negative effects on society would be disproportionately higher than the cost reduction. This can adversely affect BC ratios. Hence, in an effort to get donor funds, the size of the project is stretched beyond the point where the net benefit becomes equal to net social costs through exclusion of the negative externalities in cost calculations. This creates a socially negative consequence, namely inequity in the distribution of project benefits. In other words, those who get the benefits do not bear the costs. Since the project agencies do not earn sufficient revenue from the services they provide, adequate attention is not paid to compensating those who are adversely affected by the project. Such tendencies have also helped dam builders in inflating the net benefits of the projects.

If the donors make it mandatory for the dam builders to include the economic value of negative externality effects in the project cost, it would have the following desirable consequences. First, the agencies would be more motivated to come up with innovative designs to reduce the marginal social cost of water development. Second, they would try to improve the quality of provision of water, in order to raise the marginal value of the water. By doing this even with lower level development, the net social welfare from large dam projects could be enhanced.

Conclusions

This chapter has investigated two issues: 1) does the current criterion of height used by ICOLD for classifying large dams adequately capture the magnitude of their social and environmental impact?; and 2) are the objectives, criteria and parameters currently used to evaluate the costs and benefits of large dams sufficient to make policy choices between conventional dams and other water harvesting systems?

Analysis of data for 13,631 large dams around the world shows that the height of dam does not determine the storage volume, which, in a way, is what indicates the safety hazards posed by dams. Further analysis using data for 9,878 large dams shows that height does not determine the amount of land submerged by reservoirs. The use of such criteria results in an over-estimation of negative impacts such as displacement, leading to overreaction, from the environmental lobby, against large dams.

Analysis of data for 156 large dams in India shows that the number of people displaced by dams is a linear function of the total area submerged by them. Every one square kilometre of area submerged by large dams in India displaces around 154 people. Using the estimate of 49,660 km² of area submerged by large dams, the total population displaced by them was calculated to be 7.845 million people. As shown by the analysis, while the area submerged by dams could be an important criterion for deriving more reliable statistics about displacement, the available figures of dam-related displacement in India are gross over-estimates in an order of magnitude of eight. Therefore, the criteria for classifying dams should also include storage volume and submergence area to reflect the economic, social and environmental challenges. Apart from costs displacement and ecological degradation, the criteria for evaluating the impact of dams should include positive externalities associated with larger socio-economic and environmental benefits, such as stabilising domestic food prices, reduced carbon emissions from energy production, groundwater recharge due to return flows, and improved access to drinking water.

Water and power development agencies in poor and developing countries are not willing to transfer the additional cost of water provision, that result from the negative externalities of dam building, on the beneficiaries of dams. They fear that, with increased cost, and therefore with the increased prices users have to pay, the demand for water would come down significantly, making it difficult for them to justify large projects. Such under-estimation helps them show a higher demand for water and energy from the system, therefore enabling them to build projects of such sizes that the marginal social cost far exceeds the marginal social benefits, causing negative welfare effects on society. If the donors make it mandatory for the dam builders to build the economic value of the negative externality effects of dam building into the project cost, the net social welfare from large dam projects could be enhanced. It is argued that such an approach will also increase the pressure on the dam builders to come up with innovative systems designs that minimize the costs and raise the marginal value of water, thereby raising net social welfare.

Notes

1. According to the data cited by the author, the average live storage loss for 23 reservoirs surveyed was 0.91 per cent per annum. The actual storage in these dams, which can be diverted, would be even less.
2. Larger height meant greater foundation stresses and forces in the main body of the dam, posing geo-technical challenges. Greater storage meant greater risk for people living in the downstream. Larger spillway discharge meant greater design challenges.

References

Altinbilek, D. (2002). The role of dams in development. *International Journal of Water Resources Development, 18* (1).

Amarasinghe, U., Sharma, B. R., Aloysius, N., Scott, C., Smakhtin, V., de Fraiture, C., Sinha, A. K., & Shukla, A. K. (2004). *Spatial variation in water supply and demand across river basins of India.* Research report 83. Colombo, Sri Lanka: International Water Management Institute.

Bhalla, S., & Mookerjee, A. (2001). Big dam development: facts, figures and pending issues. *International Journal of Water Resources Development, 17* (1).

Bird, J., & Wallace, P. (2001). Dams and development: An insight to the report of the World Commission on Dams. *Irrigation and Drainage, 50.*

Biswas, A. K., & Tortajada, C. (2001) Development and large Dams: A global perspective, *International Journal of Water Resources Development, 17* (1).

Central Board of Irrigation and Power (1987). *Large dams in India, part I.* New Delhi: Central Board of Irrigation and Power, Government of India.

Central Board of Irrigation and Power (1998) *Large dams in India, part II.* New Delhi: Central Bureau for Irrigation and Power, Government of India.

Central Water Commission. (1987). *CWC guidelines for dam safety (revised).* New Delhi: Central Water Commission, Government of India.

Central Water Commission. (1994). *National register of large dams.* New Delhi: Central Water Commission, Government of India.

US Fish and Wildlife Service. Dam safety program description, definitions and standards. http://www.fws.gov/policy/361fw2.html.

Desai, S. J., & Joshi, M. B. (2008). *Narmada water plays its role: capturing initial trends from Gujarat.* Presentation, Sardar Sarovar Narmada Nigam Ltd.

Dharmadhikary, S. (2005). *Unravelling Bhakra: Assessing the temple of resurgent India,* New Delhi, India: Manthan Adhyayan Kendra.

Dhawan, B. D. (Ed.). (1990). *Big dams: Claims, counter claims.* New Delhi, India: Commonwealth Publishers.

D'Souza, D. (2002). *Narmada dammed: An enquiry into the politics of development.* New Delhi, India: Penguin Books India.

Fernandes W., Das, J., & Rao, S. (1989). Displacement and rehabilitation: An estimate of extent and projects. In W. Fernandes, & E. Ganguly Thukral (Eds.), *Development, displacement and rehabilitation* (pp. 62–88). New Delhi, India: Indian Social Institute.

Fisher, W. F. (2001). Diverting water: Revisiting Sardar Sarovar Project. *International Journal of Water Resources Development, 17* (1), 303–314.

Hira, G. S., & Khera, K. L. (2000). *Water resource management in Punjab under rice-wheat production system.* Ludhiana, India: Department of Soils, Punjab Agricultural University.

International Commission on Large Dams. (ICOLD). (2003). World Register of Dams. http://www.icoldcigb.net.

Institute for Resource Analysis and Policy. (2012). *Realistic vs. mechanistic: Analyzing the real economic and social benefits of SardarSarovar Narmada Project.* Report submitted to Sardar Sarovar Narmada Nigam Ltd., Gandhinagar, Gujarat, India.

Kumar, M. D., Ballabh, V., & Talati, J. (2000). *Augmenting or dividing: Surface water management in the water-scarce river basin of Sabarmati.* Working paper 147. Anand, Gujarat, India: Institute of Rural Management.

Kumar, M. D. (2002). *Reconciling water use and environment: Water resource management in Gujarat. Resource, problems, issues, strategies and framework for action.* Report submitted to the Gujarat Ecology Commission for the Hydrological Regime Subcomponent of the State Environmental Action Programme supported by the World Bank.

Kumar, M. D. (2003). *Food security and sustainable agriculture in India: The water management challenge.* Working paper 60. Colombo, Sri Lanka: International Water Management Institute.

Kumar, M. D. (2004). Roof water harvesting for domestic water security: Who gains and who loses? *Water International, 29* (1).

Kumar, M. D., & Singh, O. P. (2005). Virtual water in global food and water policy making: Is there a need for rethinking? *Water Resources Management, 19,* 759–789.

Kumar, M. D., Ghosh, S., Patel, A. R., Singh, O. P., & Ravindranath, R. (2006). Rainwater harvesting in India: some critical issues for basin planning and research. *Land Use and Water Resource Research, 6* (2006), 1–17.

Kumar, M. D. (2007). *Groundwater management in India: physical, institutional and policy alternatives.* New Delhi, India: Sage Publications.

Kumar, M. D., & Mudgerikar, A. (2009). *Water, human development and economic growth: Analyzing the linkages.* Proceedings of the National Conference on Hydraulics, Water Resources, Coastal and Environmental Engineering: HYDRO 2009, Pune, 17–18 December.

Kumar, M. D., Narayanamoorthy, A., Singh, O. P., Sivamohan, M. V. K., Sharma, M., & Bassi, N. (2010). *Gujarat's agricultural growth story: Exploding some myths.* Occasional paper 2–0410. Hyderabad, India: Institute for Resource Analysis and Policy.

McCully, P. (1996). *Silenced rivers: The ecology and politics of large dams.* London, UK: Zed Books.

Natural Heritage Trust. (2000). *Water resources in Australia: A summary of the National Land and Water Resources Audit's Water Resources Assessment 2000.* www.nlwra.gov.au/atlas.

Perry, C. J. (2001b). World Commission on Dams: Implications for food and irrigation. *Irrigation and Drainage, 50,* 101–107.

Ranade, R, & Kumar, M. D. (2004). Narmada water for groundwater recharge in North Gujarat: Conjunctive management in large irrigation projects. *Economic and Political Weekly, 39* (31), 3498–3503.

Rangachari, R., Sengupta, N., Iyer, R. R., Banergy, P., & Singh, S. (2000). *Large dams: India's experience, final report, prepared for the World Commission on Dams.* Cape Town, South Africa: Secretariat of World Commission on Dams.

Rosegrant, M. W., Paisner, M. S., Meijer, S., & Witcover, J. (2001). *2020 global food outlook: Trends, alternatives and choices. A 2020 vision for food, agriculture and the environment initiative.* Washington, DC, USA: International Food Policy Research Institute.

Roy, A. (1999). The greater common good. *Frontline, 16* (11) (May 22 –June 04).

Singh, O. P. (2004). *Water-intensity of North Gujarat's dairy industry: Why dairy industry should take a serious look at irrigation.* Paper presented at the second international conference of the Asia Pacific Association of Hydrology and Water Resources (APHW 2004), Singapore, June 5–9.

Singh, S. & Banerji, P. (2002). *Large dams in India: Environmental, social and economic impacts.* New Delhi, India: Indian Institute of Public Administration (IIPA).

Smakhtin, V., Revenda, C., & Döll, P. (2004). *Taking into account environmental water requirements in global-scale water resources assessments.* Comprehensive assessment report 2. Colombo, Sri Lanka: Comprehensive Assessment Secretariat.

Talati, J., & Kumar, M. D. (2005). *Quenching the thirst of Gujarat through Sardar Sarovar Project.* Paper presented at the XII world water congress, New Delhi, 22–26 November.

Thakkar, H. & Bhattacharyya, S. (2006). Reservoir siltation: Latest studies. Revealing results, a wake up call. *Dams, Rivers & People.* http://sandrp.in/dams/reservoir_siltation_in_india0906.PDF.

United Nations Development Programme. (UNDP). (2006). *Human development report.* New York, USA: United Nations.

United States National Inventory of Dams. http://crunch.tec.army.mil/nid/webpages/nid.cfm.

Verghese, B. G. (2001). Sardar Sarovar Project revalidated by Supreme Court. *International Journal of Water Resources Development, 17* (1), 79–88.

Vyas, J. (2001). Water and energy for development in Gujarat with special focus on the Sardar Sarovar Project. *International Journal of Water Resources Development, 17* (1), 37–54.

World Commission on Dams. (2000). *Dams and development: A new framework for decision making.* London, UK: Earthscan Publications Ltd.

5 Sector reforms and efficiency improvement

Pointers for canal irrigation management

Nitin Bassi, M. Dinesh Kumar and M. V. K. Sivamohan

Introduction

The role of irrigation in meeting the world's food supplies is expected to increase as developing countries are likely to expand their irrigated area from 202 million ha in 1997–99 to about 242 million ha by 2030. Most of this expansion will occur in land-scarce areas where irrigation is already critical. South Asia and East Asia will add 14 million ha each; and in land-abundant sub-Saharan Africa and Latin America, where both the need and the potential for irrigation are lower, the increase is expected to be only 2 million and 4 million ha respectively (FAO, 2002).

Given the importance of irrigation as a tool for alleviating and mitigating hunger in the developing world, it has always been a prime focus for reforms, the world over. The World Bank alone had lent around 35 billion dollars for irrigation development, which is 7 per cent of all its lending since the 1950s (Plusquellec, 1999). In spite of such huge investment, the irrigation sector continues to be caught in a vicious circle. It has been observed worldwide that lack of basic infrastructure for irrigation, poor maintenance of existing systems, and reduction of government investment in repair and rehabilitation of systems have been the major precursors for irrigation reforms (Madhav, 2007). Irrigation reforms started as early as the 1960s in Bangladesh and the USA, the 1970s in Mali, New Zealand and Colombia, and the 1980s in the Philippines, Tunisia and the Dominican Republic. The new generation reforms have taken place in Sudan and Pakistan (2000), India (late 1990s), China (2002) and, more recently, in some of the Central Asian countries. Presently more than 60 countries in the world have undergone irrigation sector reforms of one type or another (Munoz et al., 2007).

Irrigation management transfer (IMT) is one of the most important reforms in the irrigation sector. Under IMT, attempts are made to decentralize irrigation management functions and facilitate the active involvement of end users in irrigation management. This is affected through the formation of local level institutions formally called "water users associations" (WUAs). Though IMT was launched, with a big bang, as a panacea for the problems facing public irrigation, of late it is under attack from many scholars for having failed to deliver the intended benefits (van Koppen et al., 2002; Meinzen-Dick, 2007). However, some other scholars, based on limited evidence, continue to advocate IMT as the only way to ensure equity in the system-wide distribution of water (McKay and Keremane, 2006).

Field evidence relating to the impact of IMT shows mixed results. In some countries it resulted in improved system performance and an increase in the irrigated area, whereas in others fewer positive outcomes were realized. Lessons learnt from a survey of 44 IMT programmes worldwide suggest that future IMT programmes should concentrate on the following: a) WUAs and irrigation agencies need substantial capacity development; b) IMT programmes need systematic public awareness campaigns, consultations and the involvement of all key stakeholders; c) IMT should be tailor-made and flexible; and d) checks and balances should be created to ensure that WUAs act according to the members' interests (Munoz et al., 2007). IMT experiences in the Indus irrigation system of Pakistan have demonstrated that lack of role clarity between different organizations after transfer, insufficient experience and resources for water users' mobilization, lack of a democratic approach to establishing WUAs, political involvement and fear of loss of authority of government departments, have been the major factors responsible for poor progress in implementing participatory irrigation management models (Khan et al., 2007). Two action research projects in seven irrigation schemes across India, Nepal and Kyrgyzstan demonstrated that, to improve irrigation governance and water distribution by end users, provision of an appropriate legal, financial and political environment is a must (Howarth et al., 2007). Hodgson (2007), while commenting on a government of Iran/World Bank funded Alborz integrated land and water management project, also emphasized the need for proper legislation to ensure the sustainability of WUAs.

Irrigation management transfer in India

In India, various policy reforms have been carried out over the past decade in the water sector, including irrigation. This is primarily because a) water, which is becoming scarce in many regions, requires judicious management, and b) the country's surface irrigation systems are deteriorating. The problems facing Indian irrigation sector include a) declining investment in maintenance; b) low levels of system efficiency; c) poor financial working; and d) low levels of quality, reliability, and system-wide equity. Further, there is a competing demand for water, that is steadily increasing, by other sectors.

It was felt that, to improve irrigation management, it is important to involve end users/farmers in the operation and maintenance of the physical systems. The basic idea behind farmer-managed irrigation systems was to improve the overall efficiency, generate a sense of ownership among farmers and to improve the irrigation revenue recovery rate. This laid the foundation for irrigation management transfer. As a result, various state governments enacted IMT legislations. These states include Andhra Pradesh, Chattisgarh, Gujarat, Madhya Pradesh, Maharashtra, Odisha and Tamil Nadu.

Pant (2008) described the four distinct phases of IMT during the last three decades: first from 1975–85 when the emphasis was on creating outlet-based water user organizations; a second phase from 1985–90 when focus shifted to experimentation and establishment of pilot IMT projects; a third phase from the early 1990s, when few of the progressive states, such as Maharashtra, propagated the idea of turning over management of irrigation systems to the farmers; and the fourth phase which started in 1997, marking the emergence of donor funding for restructuring India's irrigation sector with IMT as a core project activity. During the third phase, the first Farmers' Management of Irrigation Systems Act was passed by the government of Andhra Pradesh in 1997.

The need for a reconsideration of IMT

Mere enactment of legislation does not ensure solutions to the problems of the country's irrigation sector (Bassi and Kumar, 2011). Over the different plan periods, the proportion of irrigation potential utilized (IPU) against that created (IPC) through surface sources does not show significant improvement (Figure 5.1). In fact, from 92.6 per cent in 1966–69, it has come down to 82.5 per cent in 2002–07. Furthermore, between the ninth (1997–02) and tenth five year plans (2002–07), there was no considerable increase in the gross irrigated area through canals. In cumulative terms, it increased from 31 million hectares in 1997–02 to only about 34 million hectares in 2002–07, an increase of only 9.7 per cent. This gain is highly insignificant considering the investments that were made on the major and medium irrigation projects during this period. From around INR 430 billion in 1997–02 (US$1 equates to INR 50), investment was increased to INR 712 billion in 2002–07, a rise of 65.6 per cent. Much less was the achievement in terms of the quality of maintenance of the conveyance systems, the timeliness in water delivery, equity in water distribution (DSC, 2003) and efficiency in fee collection. This was the case despite the growing emphasis on the involvement of end users in irrigation management.

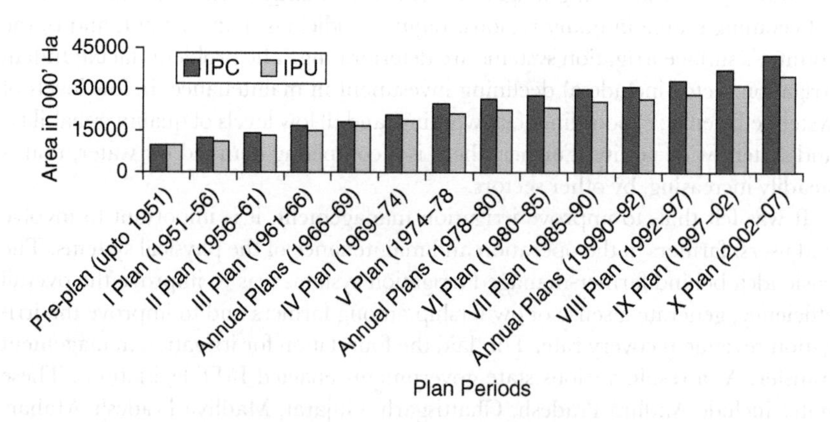

Figure 5.1 Irrigation potential created and utilized under major and medium irrigation schemes in India, by plan.

Source: Indiastat database and Central Water Commission, MoWR, Government of India.

Studies have shown that progress of IMT across states and across different types of systems has been varied (ADB, 2008) and resulted in only marginal improvement in the performance of transferred irrigation systems (Parthasarathy, 2000; van Koppen et al., 2002; Eliyathamby and Varma, 2006). Some of the reasons identified for this are: a) haste in creating farmers' organizations without any capacity building of farmers as found in Andhra Pradesh; b) transfer of systems without necessary repair and rehabilitation, work as found in Gujarat (Bassi et al., 2010) and Maharashtra (Bassi, 2008); and c) lack of appropriate legal back-up for end user organizations, as found in Punjab and West Bengal.

The researchers working on irrigation management reforms have focused on the issues involved in local institutional development (WUAs) and their effective functioning, such as the design of canal irrigation (Narain, 2008), the condition of the physical infrastructure (Mishra et al., 2011), local politics (Nikku and van der Molen, 2008), financial autonomy (Parthasarathy, 2000), capacity building of WUAs (Doraiswamy et al., 2009), member dynamics (Meinzen-Dick et al., 2002), the political and bureaucratic will to share power with farmers (Aheeyar, 2004) and the way these factors influence the success of irrigation management transfer schemes (ADB, 2008). In fact, it was suggested as early as 1994 that what was required in the reform process is not IMT, but irrigation management partnerships between government and the irrigators (Sivamohan and Scott, 1994).

Thus, scholars have mostly focused on the performance of farmers' organizations but not much on the act or policy which shaped the organizations. In order to understand the factors that lead to success or failure, it is critical to look into the formulation and implementation of the IMT reform process. The role of legislation and people who implement those laws becomes important. Acts and policies will always be effective if they are formulated and implemented as per local needs and priorities (Bassi and Kumar, 2011). Therefore it is important to investigate the following: For what purpose is the Act being designed? Who will be the stakeholders? Who will implement it? What will be the role of policy makers in its implementation? Are the reforms carried out as per the Act rationale? Often, Act formulation and implementation are considered unrelated activities, but this notion does not hold true in practice. This dichotomy can actually impact the policy outcome in significant way. This chapter highlights the IMT policy process followed in the western Indian state of Gujarat; the central Indian state of Madhya Pradesh; and the state of Maharashtra.

IMT policy process in Gujarat

By 1994, the Gujarat government's water resources department (GWRD) had launched 13 pilot IMT projects across the state. As a result of the success of some pilot projects and strong non-governmental organization (NGO) support, especially of AKRSP-I and the Development Support Centre (DSC), a Resolution (GR) was passed on 1st June 1995 which made IMT into state policy. Participation by farmers and NGOs was invited for the management of public irrigation systems. Between 1995 and 2000, some 37 orders were issued from the GWRD to facilitate IMT processes. Some of the important orders include (per DSC, 2006) a) the rehabilitation of canals prior to transfer, the WUAs to contribute 10 per cent of the costs; b) general orders for canal rehabilitation work to be first offered to the WUAs, then to the NGOs and, if both decline, to be done by government; c) WUAs willing to execute the work to be given one third of the estimated costs in advance; d) transfers of canals (pre-rehabilitation) to the willing WUAs who sign an MOU, which includes agreed estimates for the rehabilitation work to be done by the government and the physical and financial targets for completion; and e) WUAs to collect water fees and retain 50 per cent. Under the Act, the WUAs are also free to set water rates above the government rate, but not exceeding 10 per cent in any case and the additional amount can be retained by the WUAs.

During the initial stages of the reform process, capacity building work was initiated and clarification on agency reservations about IMT was facilitated by AKRSP-I and WALMI. DSC, too, played an active role in promoting IMT, by organizing series of community awareness programmes in the command areas of the Dharoi Irrigation Project, North Gujarat. Different trainings programmes for WUAs on canal rehabilitation, maintenance of records, development and enforcement of norms for water use, ensuring equity in water distribution, and regular liaision with the irrigation department were also conducted. Between 1995 and 2005, the collaborative effort between the water resources department and the DSC was successful in implementing IMT in 18 villages covering 5,100 ha. Investment made during this period, on the rehabilitation of canals, organization of the community and the operation and maintenance by WUAs amounted to INR 22.4 million (ADB, 2008). By the end of 2005, a total of 377 WUAs had been formed in Gujarat. The core tasks assigned to WUAs included a) procurement of water from the irrigation department and the distribution of it among members; b) operation and maintenance of irrigation systems under their jurisdiction; c) collection of water charges from members; d) settlement of water disputes among members; and e) maintenance of meetings records (Bassi et al., 2010).

After seeing the satisfactory working of IMT in the state, the government of Gujarat finally enacted the Gujarat Water Users Participatory Irrigation Management Act 2007. The Act added to those measures that had been passed in earlier GR orders. Major measures included an agreement to turn irrigation management by the WRD over to WUAs; and for the supply of water from the minor canals on a volumetric basis.

Some contentious government resolutions

Before enactment of the Gujarat PIM Act (2007), more than 37 GRs were passed. But there were two controversial orders. First was in 2004–05, when a GR was passed which permitted water for irrigation to be given only to those villages which had WUAs. The second was in 2005, and provided that IMT can be done without the rehabilitation of minor canals. The prime intention behind these orders was to encourage farmers' participation in irrigation management and to overcome the financial constraints of transferring minor canals with rehabilitation. However, our field experience suggests that these orders have actually resulted in the poor performance of transferred irrigation systems. Making it compulsory for villagers to form WUAs in order to receive irrigation water, without providing proper financial and institutional assistance to them, led to the creation of farmer associations only on paper. Furthermore, as seen in the command areas of the Dharoi project, farmers were not able to arrange funds to rehabilitate the transferred minor canals and thus the performance of systems did not improve. Moreover, this discouraged farmersfrim becoming members of WUAs as they feared that, despite putting their time and effort into the operation and maintenance of minor canals, they might not be able to reap the desired benefits.

Outcomes of IMT in Gujarat

Gujarat has an ultimate surface irrigation potential of about 3.3 million hectares. Of this, by 2002–07, a potential of about 2.8 million ha had been created and only 1.9 million ha were utilized. IMT is to be implemented in about 1.5 million ha of command area in the state. Additionally, in about 1.8 million ha of the command area under the Sardar Sarovar Narmada Project (SSNP), Sardar Sarovar Narmada Nigam Limited has been pursuing its own IMT model. As per the latest data available, 576 WUAs were created and around 96.7 thousand ha of area was transferred to them. But, even after 17 years of IMT, the area transferred and irrigated through WUAs' assistance is inconsequential in comparison to the total irrigated area in the state.

Furthermore, water rates for irrigation, revised in 2007, had little effect (Table 5.1). The Gujarat PIM Act states that water for irrigation has to be supplied on a volumetric basis to the WUAs, but the revised water rates were based on the area irrigated and crop type, rather than the volume of water delivered. This move was

Table 5.1 Seasonal irrigation water rates for some crops, Gujarat

Season	*Crops*	*Irrigation water rates (INR/ha)*		
		1981	*2001*	*2007*
Rabi	Wheat and mustard	110	200–240	160
	Cumin, fennel and isabgol	200	360–440	160
	Vegetables and grass	100	180–220	160
Kharif	Paddy	110	300–360	160
	Bajra	40	70–90	160
	Sorghum	40–60	70–130	160
	Maize	40	70–90	160
	Cotton and groundnut	100	180–220	160
Two seasons and Perennial	Cotton and tobacco	75–125	135–275	160
	Onion	NA	80–260	160
	Sugarcane	170–370	380–2750	300
	Banana	NA	310–2200	300

Source: Indiastat database.

in violation of the Act. Also, the gross receipt of water charges and other revenues increased only marginally from 5.6 per cent of the capital expenditure in 1995–96 to about 9 per cent in 2006–07 (CWC, 2010), in spite of the major emphasis on IMT.

Field experience

A field study carried out on eight WUAs in the Dharoi Irrigation Project, where DSC has facilitated the formation of WUAs, threw up some interesting insights into WUA performance. Performance was measured in terms of a) transparency in relation to record-keeping and WUA functioning; b) conflict resolution in terms of resolving disputes among WUA members; c) equity in relation to water distribution across various reaches of the minor canal; d) the relationship between the WUA, the irrigation department and the NGO (DSC), and also amongst farmers within a WUA; e) participation, in respect of members' attendance at various WUA meetings; and f) efficiency in respect of overall operation and maintenance of the transferred irrigation system. Five WUAs, which benefited from an adequate number of community-organizing activities, fared well on all the parameters. In comparison to this, the three WUAs for which adequate community-organizing efforts were not made performed unsatisfactorily.

Furthermore, it was found that the tail-end farmers were more inclined towards IMT than those located at the head or middle reaches. At the tail-end, more than 80 per cent of the farmers in the command area were found to be members of the WUAs, whereas at head-reach, it was only about 65 per cent. Although the highest proportion of members receiving waater for irrigation was in the head region (93 per cent against 76 and 38 per cent in middle and tail reach, respectively), distribution of water was more equitable across the tail-end WUAs (Figure 5.2). Out of the eight selected WUAs, conflicts in respect of water distribution were seen in four. In all these WUAs, the conflicts were mainly between farmers belonging to upper and lower castes. In the remaining four WUAs no conflict

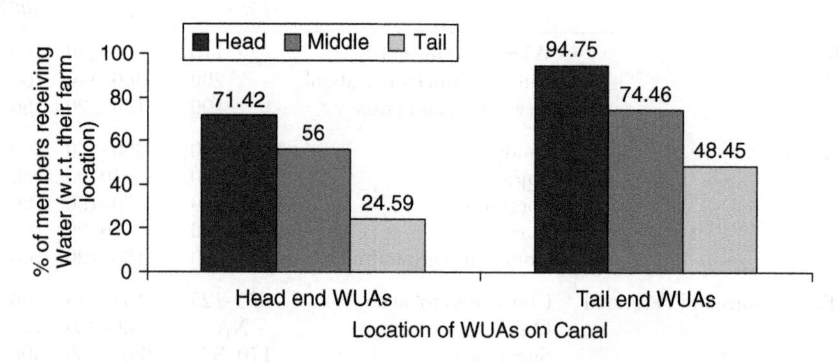

Figure 5.2 Equity in canal water distribution, Dharoi irrigation project, Gujarat.

Source: authors' own analysis using the primary data.

was seen. The farmers located at the tail-end helped in the maintenance of minor canals and those at the head-end allowed water to reach the tail-end. These are the WUAs where community-organization work was adequate. It should also be noted that, in seven of these selected WUAs, minor canals were transferred to end users without any rehabilitation. Though there were some improvements in water delivery and the conditions of canals, operation of these canals remained largely unsatisfactory. This situation led to insecurity among many farmers, in respect of their ability to access water, which forced them to remain outside the WUAs. As a result of the high cost of canal maintenance, two WUAs were running losses. Thus, though there were a few successful cases of WUAs operation, the disjointed reform process had led to a failure of IMT to provide the expected benefits.

The IMT process in Madhya Pradesh

Madhya Pradesh surface irrigation systems were heavily under-utilized due to system deficiencies, deferred maintenance, a paucity of funds to meet O&M cost and the non-involvement of farmers in irrigation management (Agrawal, 2005; Pandey, 2006). To improve the overall situation, institutional reforms were initiated by the Madhya Pradesh government in the form of the creation of irrigation Panchayats (in early 1984–85) and farmer management committees (1994–95). But, in the absence of a proper legislative framework or clear-cut governance mechanism, these attempts produced little success (Bassi and Kumar, 2011).

Based on past experience, it was realized that an important prerequisite for involving farmers in irrigation management was to create an enabling legal framework. A congenial environment was created in the state by discussions and interactions between beneficiary farmers and the agency representatives. The irrigation department had the responsibility for providing suggestions on the formulation of the IMT Act by looking at the procedures followed worldwide. Various officials from the water resources department, all line departments and academics also provided input and suggestions at the various stages of formulation of the legislation. Meetings and discussions were held with progressive farmers about the need and importance of IMT. Interestingly, NGOs were not involved in these initial stages of discourse. Finally, after due consideration of the views expressed by participants to the consultative process, the Madhya Pradesh Participatory Irrigation Management Act was passed and brought into force in 1999. The dual purpose of the Act was to improve system condition and to involve end users in irrigation management (Bassi and Kumar, 2011).

During the initial stages of the implementation of IMT, all of the financial support required for carrying out regular functions, including maintenance and administration, was provided to the WUAs by the state government. However, a major impetus in the implementation of IMT came in 2002, when the state received financial aid from the Indo-Canada Environment Facility (ICEF) to speed up the process of implementation of IMT. This support was for a period of four-and-a-half years, to assist in the physical work on seven of the transferred irrigation systems and capacity building of both officials from the water resources department officials and farmers themselves.

Reforms to implement the Act

Modeled on the Farmers' Management of Irrigation Systems Act in Andhra Pradesh, Madhya Pradesh formulated an Act dealing with irrigation reform. Under the Act, three tiers of farmer organizations were formed: WUAs at the minor canal level; distributory committees at the distributory canal level; and project committees at the level of the irrigation scheme. By default, all of the farmers having irrigable land in the jurisdiction of the water users' association became its members. By 2000–01, management committees of 1,470 water users' associations, 90 distributory committees and 57 project committees had been formed through an election, with each one having a five-year term. Major responsibilities assigned to the farmers' organizations included a) the preparation and implementation of a warabandhi schedule for each irrigation season; b) preparation of a plan for, and the carrying out of maintenance of the irrigation system in its area of operation; c) monitoring of the flow of water from the outlets; d) the resolution of disputes arising between members and water users in its area of operation; e) maintenance of accounts; f) assistance in the conduct of elections; and g) conduct of various meetings.

Major reforms were carried out at the administrative and governance level. For efficient monitoring and evaluation of IMT activities, in 2000 a separate PIM directorate was formed within the water resources department. The directorate was also made the core agency for organizing and conducting various training for farmers' organization members and water resource department functionaries involved in implementation. In addition to this, one person from the office of the chief engineer and one from the office of the executive engineer were nominated as core officers of PIM. The main responsibility of the core officers is to collect information regarding various WUA activities and to compile a progress report.

The district collector was empowered to delineate those portions of the command areas, from within each one of the irrigation systems in the district, that were to be transferred to the WUAs. This delineation was done on hydraulic system considerations. Similarly, delineation of command areas for distributory committees and project committees was done by the state government in consultation with the district collector. The district collector was also made responsible for overseeing the monthly progress of each WUA.

In accordance with the IMT Act, competent authorities were deputed to different farmer organizations. The main responsibility of these authorities was to coordinate activities between the state government departments and the farmers' organizations. Sub-engineers were made responsible for assisting WUAs in preparing a detailed list of the work to be undertaken by them and in the preparation of cost estimates for the same. However, the power to give technical clearance for the works to be undertaken by the WUAs vested in higher authorities. Furthermore, a sub-engineer was appointed as an *ex officio* member of the WUA. One staff member from the administrative cadre of the water resources department and one from the agriculture department were also made *ex officio* members of the WUA. For the collection of water charges, a lower-level government official called the *amin* was designated.

Irrigation water rates, too, were revised several times after the IMT Act came into being. In 2005, the state government decided to charge for irrigation water on the basis of the amount of watering of crops and, in the process, put more financial pressure on farmers. The charges had previously been based on irrigated area.

Outcomes of IMT in Madhya Pradesh

Out of the ultimate surface irrigation potential of 2.45 million ha, by 2006–07, 2.7 million ha had been created, but only 1.68 million ha utilized. Furthermore, only 1.7 million ha of the irrigable command area had been transferred to the farmers' organizations. Also, the net area irrigated by canals in the state increased only marginally, from 0.8 million ha in 2000–01 to only 1.05 million ha in 2007–08. A strong correlation was seen between the irrigated area and the revenue collected. In 2003–04 and 2004–05, when there was an increase in the net canal irrigated area as compared to previous years, the revenue collected also went up considerably. But, overall, irrigation charge recovery remained low, averaging around 56 per cent from 1998–99 to 2008–09 (Figure 5.3). Also, the gross receipt on account of water charges and other revenues had decreased substantially, from 11.5 per cent of the capital expenditure in 1998–99 to about 2.8 per cent in 2006–07 (CWC, 2010).

In terms of the number of farmers' organizations formed and the irrigable command area transferred to them, no significant increase was observed between the two election terms of water users (2000–01 and 2006). The number of farmers' organizations increased by only 13 per cent, and the irrigable command area

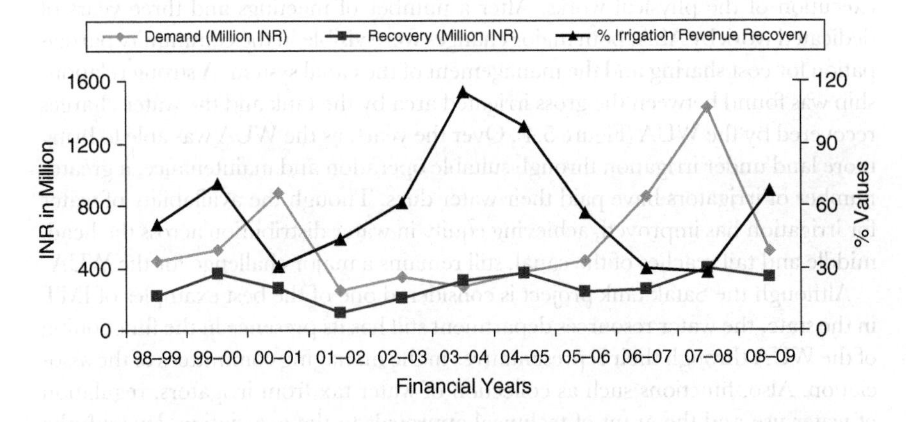

Figure 5.3 Irrigation revenue recovery by year, Madhya Pradesh.

Source: Water Resources Department, Madhya Pradesh.

under these organizations by only 12.5 per cent. This was quite expected, as the focus of the PIM project during this phase was on only seven irrigation schemes.

Insight into the IMT in Satak tank project, Khargone district

Satak tank project is a medium irrigation scheme, constructed during 1955–66 on the Satak river in the Narmada river basin. The project has 2,706 hectares of culturable command area, out of which nearly 1,800 ha are irrigable. The tank has a total distribution network of 53 km, which covers 17 villages comprising 1,750 water user families. The distribution network consists of one main canal, one distributory and 13 minor canals.

Under the IMT process, only one WUA was formed in the entire command area. The canal distribution network was transferred to WUA without any necessary repair and rehabilitation works. Until the early part of 2003, the major role of the WUA was limited to annual maintenance of the canal structure. In the second half of 2003, the Satak tank project was included in the ICEF project and INR 12.8 million was assigned for the renovation of the whole canal network. A notable clause in the project was related to the total expenditure on the execution, which was to be shared by the donor agency, the state government and farmers in a ratio of 50:20:30. However, because of the farmers' inability to contribute their stipulated share of 30 per cent, the proportion was later changed to 50:30:20 and then again to 60:30:10. It was one of the few projects where a community made a full contribution towards the total expenditure. However, under the state IMT Act, all of the money was deposited with the water resources department and its release was subjected to the technical clearance of physical works by the competent authority.

Under the ICEF project, a NGO named ASA was involved as a facilitating agency to a) promote local institution-building at grassroots level; b) motivate farmers to pay their contribution; and c) to provide guidance to the farmers for the execution of the physical works. After a number of meetings and three years of dedicated work by ASA, both major changes were visible in the community participation for cost sharing and the management of the canal system. A strong relationship was found between the gross irrigated area by the tank and the water charges recovered by the WUA (Figure 5.4). Over the years, as the WUA was able to bring more land under irrigation through suitable operation and maintenance, a greater number of irrigators have paid their water dues. Though the availability of water for irrigation has improved, achieving equity in water distribution across the head, middle and tail reaches of the canal, still remains a major challenge for the WUA.

Although the Satak tank project is considered one of the best examples of IMT in the state, the water resources department still has its presence in the functioning of the WUA through their representatives in the managing committees of the association. Also, functions such as collection of water tax from irrigators, regulation of water use and the grant of technical approvals to the association, lay with the water resources department. The role of the WUA was limited to maintenance of the canal system and motivating farmers to pay their water charges on time.

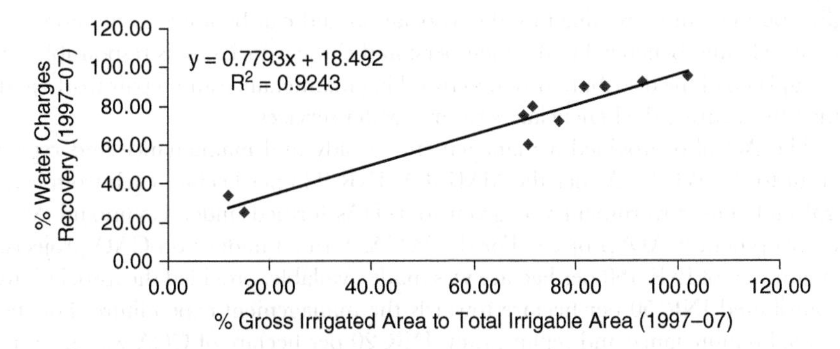

Figure 5.4 Relationship between gross irrigated area and water charges recovery, Satak tank project.

Source: authors' own analysis using the primary data.

After the completion of work supported through the ICEF project, ASA moved out of the project area. During this phase, elections to select representatives for the second term of the WUA were held and new office-bearers have taken over the functioning of the association. But, in the absence of any financial and institutional support, the new management committee members of the association were uncertain about their role and what was expected of them in terms of improvement in the performance of the transferred irrigation system.

IMT policy process in Maharashtra

Maharashtra has a rich tradition of farmer involvement in the management of irrigation systems: be it the phad system or shejpali, farmers in the state have been collectively instrumental in securing water for irrigation. The first initiative by the government of Maharashtra in moving towards IMT came with the Cooperative Water Users' Association Guidelines (1994) whereby the irrigation department adopted a policy to a) create water users' associations at minor canals; b) transfer operation and maintenance responsibility for the minor and smaller channels to WUAs; c) allocate water to WUAs through yearly agreements; and d) charge WUAs for irrigation water supplied on volumetric basis (Naik and Kalro, 1998).

Since then, there have been continuous changes in the guidelines for involving farmers in irrigation management. The state government passed the Maharashtra Management of Irrigation Systems by Farmers Act (MMISFA) 2005. Under this Act, WUAs were entrusted with following rights and responsibilities: a) after receiving water on a volumetric basis, the right of internal distribution of water among the farmers rests with the association; b) WUAs can levy different water charges for members and non-members – the only restriction is that the rates levied on non-members should not be more than 130 per cent of those charges to the members; c) if the WUA saves water from their allotted quota for the rabi (winter) season, the saved quantity of water can be used by the association for irrigating crops during summer season; d) profit accrued to the association through water distribution can be retained by the association and can be used for undertaking other schemes beneficial to the members; and e) the association is responsible for the upkeep of the distribution system by taking maintenance and repair works and thereby ensuring the beneficiaries receive better services.

The Act also provided a management subsidy and maintenance and repair grant to the WUAs. As per the MMISFA, INR 225 per hectare each from central and state government was given to WUAs formed under Command Area Development (CAD) projects. For the WUAs formed under non-CAD projects, an amount of INR 450 per hectare was made available, provided the associations contributed INR 50 per hectare towards the management expenditure. For the annual maintenance and repair grant, INR 20 per hectare of CCA was given to WUAs before MMISFA; this was increased to INR 60 per hectare after implementation of MMISFA. However, the grant will taper from sixth year onwards and will be stopped once WUA completes 10 years.

In order to facilitate the smooth transfer of irrigation management functions to WUAs, various capacity-building activities were organized. WALMI-Aurangabad was entrusted with the responsibility of imparting training to the office-bearers and members of the WUAs, canal operators, and officers from the irrigation, agriculture and co-operative departments. By 2006, 1,127 functional WUAs covering a cultivable command area of 0.3 million ha were formed.

The role of NGOs in promoting IMT

Right from the formation of the first Irrigation Cooperative Society (Datta), formed in 1989 in the Mula irrigation command area in Ahmednagar district, NGOs have played a significant role in facilitating and advising the Maharashtra government on IMT. Following the success of the Datta Society, a Pune-based NGO, SOPPECOM, tried to make a federation of 14 WUAs. Similarly, in the Waghad irrigation system in Nashik district, all the WUAs were established with the motivation of a social organization called Samaj Parivartan Kendra (SPK) and its creator, Late Bapu Upadhyaya. While the total number of WUAs set up by NGOs may be small, they have demonstrated the important role they have played in developing IMT in the state (ADB, 2008).

Outcomes of IMT in Maharashtra

Out of an ultimate surface irrigation potential of about 5.3 million ha in Maharashtra, by 2002–07, a potential of 4.6 million ha was created but only 3 million ha were utilized. As per the state Act, IMT has to be implemented in the entire command areas of surface irrigated projects. By April 2010, about 2,815 WUAs had been created and only 1.1 million ha of irrigated area had been transferred to them. Thus, as with Gujarat and Madhya Pradesh, the area taken up by WUAs is insignificant. Also, most of the minor canals were transferred to WUAs with only selective repairs. As a result, there was no significant increase in the irrigated area because of IMT.

However, the ratio of recovery of irrigation water charges to the cost of providing the service for the major and medium irrigation schemes has shown substantial improvement (GOM, 2011). From 58 per cent in 2004–05, it has increased to 98.5 per cent in 2009–10. But this improvement is because of a decrease in the amount of money spent on operation and maintenance of the systems, rather than any improvement in recovery of assessed water charges (termed as assessment recovery ratio), which increased only marginally from 48 per cent in 2004–05 to 55 per cent in 2009–10 (Figure 5.5). Also, the gross receipt on account of water charges and other revenues has increased only marginally, from 7.2 per cent of the capital expenditure in 2000–01 to about 8.8 per cent in 2006–07 (CWC, 2010).

While the Act affirms that water needs to be supplied to WUAs on a volumetric basis, charges for irrigation water services for the year 2008–09 were imposed based on the area irrigated under different crops. Thus it is clear that financial reforms were not carried out as per the guidelines of the state IMT Act. But, in spite of poor performance of IMT in terms of the area transferred to WUAs; the increase in amount of area irrigated through canals; and the recovery of assessed irrigation water charges, some successful stories on farmers' managed irrigation systems have emerged from the state.

Field observations on Ozar WUA, in the Nashik district

The WUA at the village of Ozar, which is around 25 km from the district headquarters, is located at the tail-end of the Waghad irrigation project. Waghad dam is a major dam (under the upper Godavari project) constructed in 1978 on the river Kolwan, which is a tributary of the Kadwa river. The right and left bank canals of the dam are designed to irrigate about 5,100 hectares and 1,650 hectares of land area respectively.

Ozar WUA was formed with the tireless and unending efforts of SPK. For organizational purposes, and to make the farmers understand the importance of

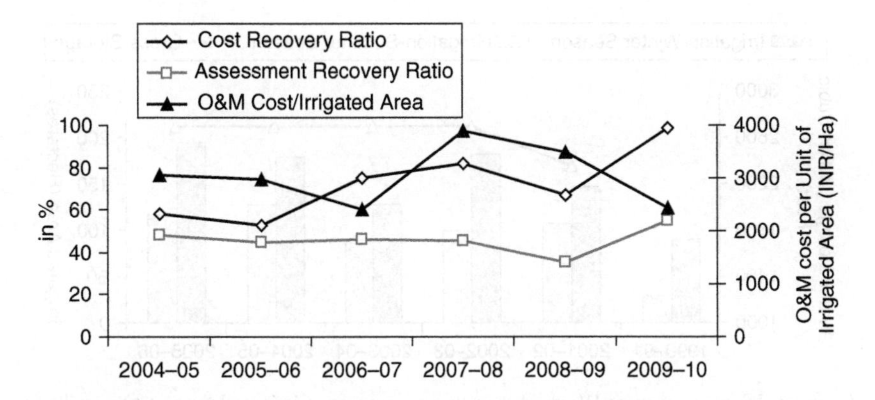

Figure 5.5 Irrigation water charges recovery and O&M cost, Maharashtra.

Source: Government of Maharashtra, 2011.

IMT, SPK organized a number of meetings. These efforts became fruitful when the association was finally registered in 1991 under the Cooperative Society Act[1] and started its operation on the transferred irrigation system. Due to a paucity of funds within the water resources department, the irrigation system was handed over to WUA with only selective repairing of the minors.

The performance of the transferred irrigation system improved after the formation of Ozar WUA. Though the gross water storage of the reservoir remained at 2,700 million cubic feet between 2002–03 and 2005–06, the WUA was able to bring a larger area under irrigation (Figure 5.6). The net area irrigated during the winter crop season increased from 85 ha in 1989–90 to more than 190 ha in 2005–06. The increase in irrigated area during the summer season was more remarkable, from only 1 ha in 1989–90 to about 116 ha in 2005–06. This increase in irrigated area was the result of improved management and optimum utilization of water by the WUAs. Farmers could get water during summer months by saving from the sanctioned quota of water during the winter season. Most of the farmers were found to be irrigating grapes during the summer season.

On the financial front, the performance of Ozar WUA was found to be satisfactory. During the period from 2000–01 to 2005–06, WUA reported losses in three years (Figure 5.7). The cost of maintenance of minor canals had increased substantially during the past few years. From INR 23,520 in 2003, it had almost doubled to INR 42,057 in 2005. Also, the total water charge paid to the state irrigation department had increased from INR 62,000 in 2003 to almost INR 160,000 in 2005. Importantly, irrigation charges recovery rate for the period between 2000 and 2006 was found to be constant (above 90 per cent). Thus, it is likely that the increasing expenditure on the maintenance of minor canals might have caused losses to the WUA. A similar financial problem was found with other successful WUAs in the Ahmednagar, Akola and Nanded districts of the state (Bassi, 2008).

Overall, Ozar WUA was found to be performing well, and looks stable. However, minor canals which were transferred to them required repairs at several

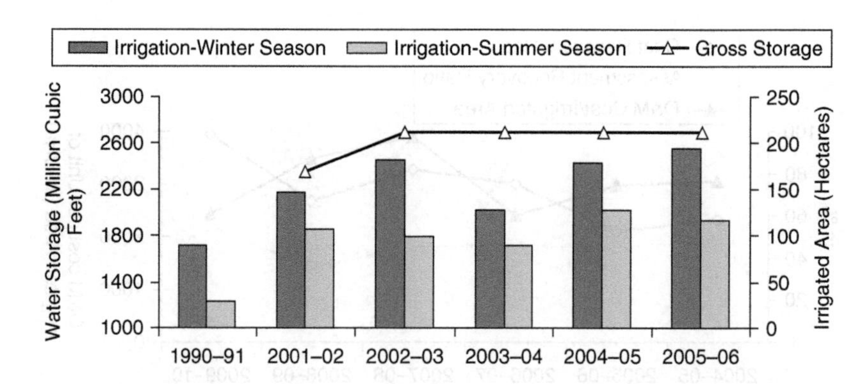

Figure 5.6 Water storage at Waghad dam reservoir and net irrigated area at Ozar village, Maharashtra.

Source: authors' own analysis using primary data.

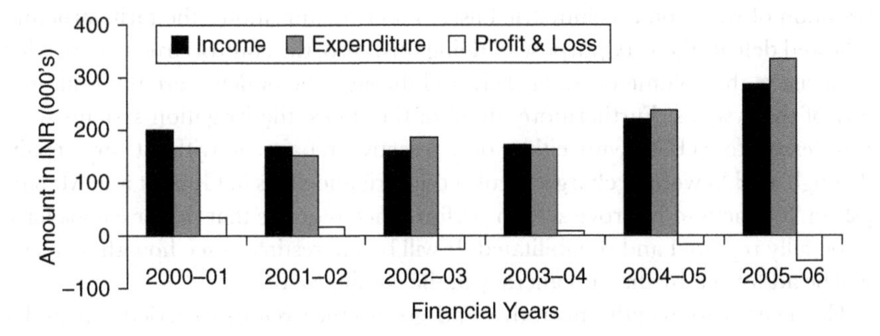

Figure 5.7 Financial performance of Ozar WUA, in Nashik district, Maharashtra.

Source: authors' own analysis using primary data.

places. Also, in view the state government decision that WUAs shall be provided with water based on the volumetric basis, the repair work will need to be given top priority. Furthermore, the volumetric supply of water will yield the desired benefit of judicious use of irrigation water only if the conveyance system is put to better condition. But the major constraint for carrying out such mass repair and rehabilitation of minors is the lack of funds with the state government.

Issues in implementation of IMT

Analyses of the IMT process followed in three states of India suggest that an incremental model of policy formulation was followed in all of the states. Under this approach, only a small number of alternatives for dealing with the problem are considered and, finally, an option is selected which differs only marginally from the existing policy (Sutton, 1999). In response to the problem of poor irrigation system performance, the states decided to implement some policy reforms. The IMT Acts were drafted with some modifications to the international models of IMT, in order to reflect the state socio-political environment, the nature of hydraulic systems, investment need and agriculture patterns.

In Gujarat, NGOs had a powerful influence on the policy making process. In Madhya Pradesh, experts from the water resources department played the most important role, whereas, in Maharashtra, both the irrigation department and the NGOs were involved. These different institutions were involved in the IMT process worldwide and finally guided the IMT policy drafting for their respective states. But, in all the three states, there were only a few meetings undertaken with selected individuals. It can be inferred that the policy formulation process did not pay attention to the perception of the majority of the real stakeholders, i.e. the irrigators and others. Thus, in all the states, a top-down approach was followed for shaping the IMT Act.

The reforms carried out under the Acts are inadequate to yield the intended outcomes. The financial reforms undertaken in Gujarat and Maharashtra do not encourage judicious use of available water. Though the Acts clearly mention distribution of water on a volumetric basis to user organizations, the tariff structure followed defeats the very purpose of doing this. Also, the control structures needed to measure the volume of water delivered through the outlets were not built into any of the systems. Furthermore, in all of the states, the irrigation systems were transferred to WUAs with either only selective repairs or without any repair. Though WUAs were in charge of collecting irrigation fees in Gujarat and Maharashtra, for them to improve system performance requires that minor canals have to be fully repaired and rehabilitated. It will be interesting to see how these issues will be addressed by the bureaucracy in the coming years.

However, various administrative and governance reforms carried out under the IMT Act in Madhya Pradesh suggested the involvement of water resources officials in the workings of WUAs. Discussions with officials suggested that they wanted to gradually transfer all irrigation management functions to the associations. These reforms hint at the narrow objectives of state governments, namely of reducing their expenditure on system maintenance and making irrigators bear the burden of collecting irrigation water charges and carrying out necessary maintenance of the systems, notwithstanding that they had been transferred without proper repairs.

Furthermore, over-dependence on donors to fund large-scale IMT implementation has not helped the creation of sustainable institutional models. This is mainly because ultimately the aid will stop and, once it stops, WUAs are left without any vision for the future. Thus, in the implementation of community welfare programmes, there is a need to craft some rules and procedures which keep the newly established system sustained even after the aid stops.

Capacity building is also an integral part of the IMT implementation process. As seen in some of the successful cases in Gujarat and Maharashtra, WUAs were able to function effectively (under certain limitations) with the support of social organizations such as AKRSP-I, DSC, SPK and SOPPECOM. It is also important that before the NGOs exit they identify potential leaders and undertake training and skill development of these identified individuals. Training is also required on the technical aspects of irrigation systems, so that the transferred irrigation system can be efficiently managed in future. These trained individuals would then be in a position to pass their knowledge of irrigation management on to other elected WUA members. Such a capacity-building process will take care of the situation observed in the Satak irrigation project, whereby new WUA management committee members were found to be wanting in managing the transferred irrigation system.

Alternative institutional models for other countries

The model of IMT policy process followed; government inability to generate funds within the system for implementing IMT; and isolated islands of success in promoting effective WUAs, clearly show that reforms for participatory irrigation management in India were inadequate in bringing about the desired outcomes.

Similar experiences can be found in many other developing economies in Asia (such as Bangladesh, Pakistan, Iran and Sri Lanka) and sub-Saharan Africa (such as Madagascar, Niger, Nigeria and Senegal). In light of such challenges, it is necessary to explore and experiment with other institutional models.

Kumar and Bassi (2011) suggested an innovative institutional model for promoting efficient water use (through establishing a system of volumetric water pricing and tradeable water entitlements) and the creation of self-driven users' associations in the command areas of the Sardar Sarovar Narmada Project in Gujarat. Under this model, intermediary institutions needs to be created at various levels in the hydraulic system hierarchy (i.e. from main canal to branch and distributory canals), apart from the WUAs. These intermediaries will help negotiate water transfer agreements between farmers and payments for the sale and purchase of water entitlements. Income gain from water trading should create new incentives, among those farmers below the minor outlets, to form water users' associations to carry out the larger irrigation management functions. The farmers will be able to mobilize their own financial resources for the following: a) hiring technical manpower for carrying out some of the difficult functions; b) investing in the hydraulic structures for measurement of water delivery; and c) the execution of works relating to physical system improvement. Since such institutions will be self-driven rather than external agency promoted, they are likely to sustain. That would also reduce the transaction cost of creating institutions. Such self-sustaining models of users' associations can be tried in the countries of sub-Saharan Africa which suffer from lack of financial resources, poor governance, weak public institutions, political instability and corruption.

Alternatively, the public–private partnership (PPP) mode of irrigation management, as seen in some parts of China, Senegal, Egypt and Saudi Arabia, also need to be explored. Under a PPP in Senegal, the public agency responsible for water management in the basin has set up a maintenance contracting unit, which is intended to be financially autonomous. The users' associations running the schemes can contract with this unit for operation, maintenance and management services on a payment basis. In China's Hunan province, the local government has delegated management of upstream irrigation and the drainage components of water mobilization and conveyance to strong water service corporations and the management of downstream water distribution functions to WUAs (World Bank, 2007).

PPPs may emerge as one way of bringing in much-needed management skills and funds and thereby relieving government of the fiscal and administrative burden. However, irrigation and drainage are not activities that immediately attract the private sector, particularly when they involve investment. But this will surely happen if there is a favourable policy environment and there are no political bottlenecks. In the public–private mode of irrigation systems management, farmers would eventually pay more for the services offered, but the increased burden caused by the higher irrigation cost can be offset by the increase in the net farm returns on account of the improved quality of irrigation water.

Note

1. Under the Maharashtra PIM Act 2005, all the WUAs have to be registered with the irrigation department.

References

Agrawal, K. N. (2005). *Participatory irrigation management in Madhya Pradesh.* Paper delivered at the workshop on attitude and behavior change for the ICEF-WRD-PIM project partners, Bhopal, Madhya Pradesh, India, 12–13 March.

Aheeyar, M. M. M. (2004). Measures to improve linkage between main system management and farm-level management. In C. M. Wijayaratna (Ed.), *Linking main system management for improved irrigation management* (pp. 17–28). Tokyo, Japan: Asian Productivity Organization.

Asian Development Bank. (2008). *Irrigation management transfer: Strategies and best practices.* New Delhi, India: Sage Publications.

Bassi, N. (2008). *Surface irrigation and livelihoods: Results of irrigation management transfer in Maharashtra.* Paper presented at the XIII IWRA world water congress, Montpellier, France, 1–4 September.

Bassi, N., & Kumar, M. D. (2011). Can sector reforms improve efficiency? Insight from irrigation management transfer in central India. *International Journal of Water Resources Development, 27* (4), 709–721.

Bassi, N., Rishi, P., & Choudhury, N. (2010). Institutional organizers and collective action: The case of water users' associations in Gujarat, India. *Water International, 35* (1), 18–33.

Central Water Commission. (2010). *Financial aspects of irrigation projects in India.* New Delhi, India: Central Water Commission.

Development Support Centre. (2003). *Tail-enders and others deprived in the canal water distribution.* Report prepared for the Planning Commission. Ahmedabad, India: Development Support Centre.

Development Support Centre. (2006). *PIM in Gujarat and DSC in PIM.* Ahmedabad, India: Development Support Centre.

Doraiswamy, R., Mollinga, P. P., & Gondhalekar, D. (2009). *Participatory training in canal irrigation in Andhra Pradesh: The Jala Spandana experience.* STRIVER Policy Brief, 17.

Eliyathamby, N., & Varma, S. (2006). *Improving performance and financial viability of irrigation systems in India and China.* IWMI Water Policy Briefing 20. Colombo, Sri Lanka: International Water Management Institute.

Food and Agriculture Organization. (2002). *World agriculture: Towards 2015/2030.* Rome, Italy: Food and Agriculture Organization.

Government of Maharashtra. (2011). *Report on benchmarking of irrigation projects in Maharashtra state: 2009–10.* Maharashtra, India: Water Resources Department, Government of Maharashtra.

Hodgson, S. (2007). *Legislation for sustainable water user associations.* FAO legal papers online 69, Rome, Italy: Food and Agriculture Organization.

Howarth, S., Nott, G., Parajuli, U., & Dzhailobayev, N. (2007). *Irrigation, governance and water access: Getting better results for the poor.* Paper presented at the 4th Asian regional conference and 10th international seminar on participatory irrigation management, Tehran, Iran, 2–5 May.

Khan, A. H., Gill, M. A., & Nazeer, A. (2007). *Participatory irrigation management in Pakistan: opportunities, experiences and constraints.* Paper presented at the 4th Asian regional conference and 10th international seminar on participatory irrigation management, Tehran, Iran, 2–5 May.

Kumar, M. D., & Bassi, N. (2011). Maximizing the social and economic returns from Sardar Sarovar project: Thinking beyond the convention. In R. Parthasarathy & R. Dholakia (Eds.), *Sardar Sarovar Project on the River Narmada: Impacts so far and ways forward* (pp. 747–776). New Delhi, India: Concept Publishing.

Madhav, R. (2007). *Irrigation reforms in Andhra Pradesh: Whither the trajectory of legal changes?* Working paper 4. Geneva, Switzerland: International Environmental Law Research Centre.

McKay, J., & Keremane, G. B. (2006). Farmers' perception on self created water management rules in a pioneer scheme: The mula irrigation scheme, India. *Irrigation and Drainage Systems, 20* (2–3), 205–223.

Meinzen-Dick, R. (2007). Beyond panaceas in water institutions. *PNAS, 104* (39), 15200–15205.

Meinzen-Dick, R., Raju, K. V., & Gulati, A. (2002). What affects organization and collective action for managing resources? Evidence from canal irrigation systems in India. *World Development, 30* (4), 649–666.

Mishra, A., Ghosh, S., Nanda, P., & Kumar, A. (2011). Assessing the impact of rehabilitation and irrigation management transfer in minor irrigation projects in Orissa, India: A case study. *Irrigation and Drainage, 60* (1), 42–56.

Munoz, G., Garces-Restrepo, C., Vermillion, D. L., Renault, D., & Samad, M. (2007). *Irrigation management transfer: Worldwide efforts and results.* Paper presented at the 4th Asian regional conference and 10th international seminar on participatory irrigation management, Tehran, Iran, 2–5 May.

Naik, G., & Kalro, A. H. (1998). *Two case studies on the role of water users' associations in irrigation management in Maharashtra, India.* Washington, DC, USA: World Bank.

Narain, V. (2008). Reform in Indian canal irrigation: Does technology matter? *Water International, 33* (1), 33–42.

Nikku, B. R., & van der Molen, I. (2008). Conflict, resistance and alliances in a multi-governance setting: Reshaping realities in the Andhra Pradesh irrigation reforms. *Energy and Environment, 19* (6), 861–875.

Pandey, A. (2006). *Ethnography of participatory irrigation management in Vidisha district of Madhya Pradesh.* Bhopal, India: Indian Institute of Forest Management.

Pant, N. (2008). Key issues in participatory irrigation management. In M. D. Kumar (Ed.), *Managing water in the face of growing scarcity, inequity and declining returns: Exploring fresh approaches* (pp. 540–557). Hyderabad, India: International Water Management Institute.

Parthasarathy, R. (2000). Participatory irrigation management programme in Gujarat: Institutional and financial issues. *Economic and Political Weekly, 35* (35–36), 3147–3154.

Plusquellec, H. (1999). The role of the World Bank and new opportunities. In Food and Agriculture Organization (Ed.), *Modernizations of irrigation system operations* (pp. 13–19). Bangkok, Thailand: Food and Agriculture Organization.

Plusquellec, H. (2002). *How design, management and policy affect the performance of irrigation projects: Emerging modernization procedures and design standards.* Bangkok, Thailand: Food and Agriculture Organization.

Sivamohan, M. V. K., & Scott, C. A. (Eds.). (1994). *India: Irrigation management partnerships.* Hyderabad, India: Booklinks.

Sutton, R. (1999). *The policy process: An overview.* Working paper 118. London, UK: Overseas Development Institute.

van Koppen, B., Parthasarathy, R., & Safiliou, C. (2002). *Poverty dimensions of irrigation management transfer in large-scale canal irrigation in Andhra Pradesh and Gujarat, India.* Research report 61. Colombo, Sri Lanka: International Water Management Institute.

World Bank. (2007). *Emerging public-private partnerships in irrigation development and management.* Water Sector Board discussion paper series 10. Washington, DC, USA: World Bank.

6 Rebuilding traditional water harnessing systems for livelihood enhancement in arid western Rajasthan

Nitin Bassi and V. Niranjan

Introduction

Western Rajasthan is characterized by low rainfall and high aridity (Goyal, 2010). This is compounded by extremely high inter-annual variability in annual rainfalls and rainy days. The years of deficit rainfall, characterized by fewer rainy days with long dry spells and higher aridity, see hydrological droughts leading to a severe shortage of water, not only for irrigation, but also for the basic survival needs of human and animal population. But wet years produce excessively high runoffs, often causing flash floods. Traditional runoff farming systems, such as khadins, can store part of this excess runoff, not only on the surface, but also in the soil profile, to enable good production of kharif and winter crops. Similarly, the village ponds had a prominent role in domestic water supply provisions in the villages of arid western Rajasthan.

But, with population growth, changing lifestyles, and the consequent exponential increase in water demands for irrigation and domestic uses, communities started depending on groundwater, which used to serve as a buffer in years of drought, to grow irrigated crops and supply water for livestock and human consumption. While groundwater resources started depleting due to over-draft, the management of traditional systems was largely ignored. Thus, in spite of the social and environmental benefits these traditional water bodies used to provide, many had fallen into disuse over the years, due to lack of proper maintenance and the simultaneous advent of modern superior water supply systems which ensured high reliability and better quality.

During the last two decades, traditional water harvesting systems such as the khadins and village ponds in Rajasthan have captured the imagination of ethnographers, water resource scholars, water management professionals and development organizations alike, with the result that initiatives to revive them have come from many quarters, including government and development organizations. The Central Arid Zone Research Institute (CAZRI), Jodhpur, has played an important role in the revival of many such traditional water harvesting systems across western Rajasthan. The Ambuja Cement Foundation (ACF), a philanthropic organization, is actively involved in such attempts in the districts of Pali and Nagaur. While a large volume of folklore and popular literature show that the benefits of reviving these traditional water bodies, there has been no systematic effort to analyze the impact of these interventions based on actual field data. This study assesses the hydrological and socio-economic impact of interventions carried out by ACF in the district of Pali and Nagaur in western Rajasthan.

Western Rajasthan: physical environment

Western Rajasthan has an arid to semi-arid climate. The region is characterized by low and erratic distribution of rainfall, extremes of diurnal and annual temperatures, low humidity and high wind velocity. With average annual rainfall of only 317 mm, this region is the most arid part of the country (Goyal, 2010). Owing to these extreme climatic conditions, the region experiences an average potential evapo-transpiration of more than 2,000 mm per year (Rao, 2009), a negative water balance and acute water deficit (Narain et al., 2005).

The mean annual rainfall in western Rajasthan ranges from less than 100 mm in the north-west part of Jaisalmer; 200 to 300 mm in the Ganganagar, Bikaner and Barmer regions; 300 to 400 mm in the Nagaur, Jodhpur, Churu and Jalore regions; and more than 400 mm in the Sikar, Jhunjhunu and Pali regions. More than 85 per cent of the total annual rainfall is during the south-west monsoon season (July to September). There is a high inter-annual variability in rainfall, as reflected in its high degree of coefficient of variation (around 45 per cent for Pali and 54 per cent for Nagaur). Recorded annual rainfall data (source: Rajasthan water resources department) from the Jaitaran block, in the Pali district (from where the khadin and village pond were selected) show similar trends (Figure 6.1). During last 52 years, the mean annual rainfall in Jaitaran was 407.9 mm, with highest annual rainfall of 872.5 mm recorded in 1979 and lowest of 122.7 mm recorded in 1972. The rainfall variability, expressed in terms of coefficient of variation, is high (44.77 per cent). The last five years (2005–09) have been even more parched, with mean annual rainfall of only 383.4 mm. There is a definite trend in annual rainfall, with a 20-year cycle consisting of wet and dry years, as shown by moving averages, but the peak values are reducing over the years.

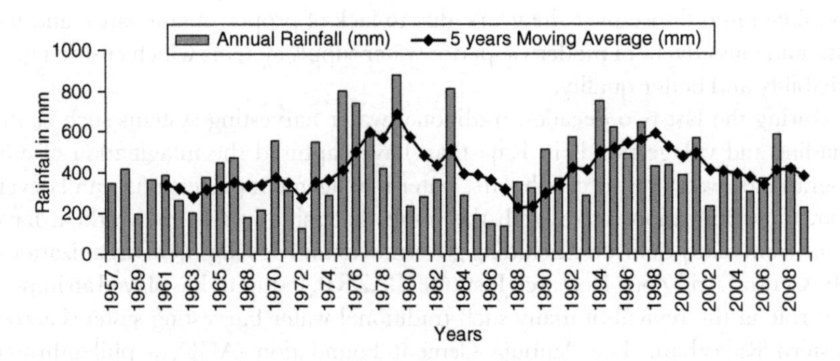

Figure 6.1 Annual rainfall and moving average (1957–2009), Jaitaran, Pali.

Source: Water Resources Department, Government of Rajasthan.

Furthermore, the years of high rainfall are also characterized by a large number of rainy days. There is a strong correlation between the quantum of annual rainfall and the number of rainy days in the year (Figure 6.2). The R square value for Jaitaran was 0.54. For Nagaur (the district from which the other village pond was selected), the mean annual rainfall was even lower (385.67 mm, based on the data record of the past 52 years) with the highest annual rainfall of 1,259 mm in 1975 and the lowest of 110 mm in 2009. With a coefficient of variation of 54.38 per cent, the region, too, shows very high rainfall variability.

In western Rajasthan, solar radiation is high throughout the year. During winter, it varies between 15.12 and 17.71 MJ/m^2/day and in summer months the values range from 22.79 to 26.50 MJ/m^2/day. The region experiences extreme air and soil temperatures, which considerably increase the water required for vegetation. Relative humidity in the region is low during summer and winter months but gradually increases to around 80 per cent by monsoon. This low humidity, combined with strong wind, leads to evaporation loss that is greater than the energy actually available through solar radiation. The annual potential evapotranspiration ranges from 1,400 mm/year in the eastern part to more than 2,000 mm/year in the western part of western Rajasthan (Rao, 2009). Soil strata in these regions are alluvial with limestone and allied sedimentary rocks.

All three hydro-geological formations, i.e. unconsolidated sediments, semi-consolidated sediments and consolidated rocks, exist in western Rajasthan. Out of these, unconsolidated formations are dominant. The alluvial deposits are confined to the Barmer, Jalore andJodhpur districts, and consist of sand, clay, gravel and cobbles. Semi-consolidated formations, which include sandstones and limestone, cover the Jaisalmer and Barmer districts. In the Pali region, groundwater occurs in a single and thin zone of saturation which extends through contiguous bodies of thin alluvium, and of igneous, metamorphic and consolidated sedimentary rocks. All of the groundwater that occurs in the rocks of this region originates from rainfall. Permeable rocks are present in the space between the land surface and the water table, making the groundwater exist in a sort of a reservoir and causing the water table to rise and fall freely as water is added to, or discharged from, the

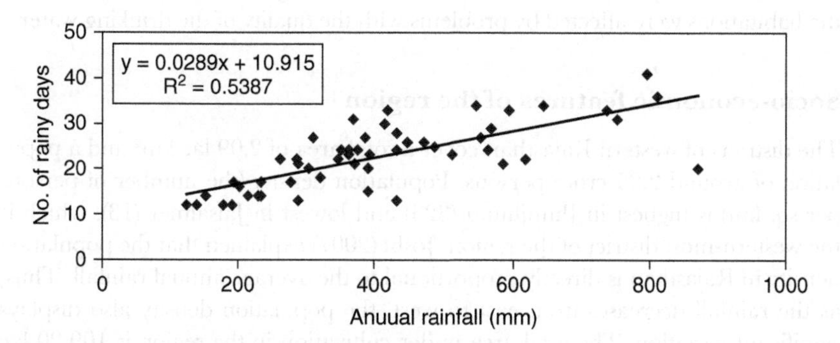

Figure 6.2 Relation between annual rainfall and rainy days.

Source: authors' own estimate of the secondary data.

rocks. Thus, the water table rises during or after heavy rainfall, and falls during the dry season (Taylor et al., 1955).

The Luni river basin covers a few of the districts, including Pali, Jalore, part of Jodhpur and Barmer in this western region of Rajasthan. The remaining part of western Rajasthanis covered by endorheic or internal drainage basins. Luni is a closed drainage basin that does not drain the water into any natural surface water sink, such as an ocean or lake. Precipitation that falls on such basins leaves the drainage system by evaporation or gets lost to the salt sinks (for more details on the concept of closed basins, please refer to Seckler, 1996 and Kumar, 2010). As per the water scarcity index (Falkenmark et al., 1989), both Luni and the internal drainage basin suffer from water scarcity (a per capita water availability of less than 1,000 m^3/annum).

Groundwater development in western Rajasthan exhibits alarming trends. Out of the 11 districts in the region, groundwater resources are over-exploited in seven, critical in two, and semi-critical in one district. With a negative groundwater balance of 1,264.19 MCM and a stage of groundwater development of nearly 140 per cent, the region has limited scope for further groundwater development except in few blocks. The stage of groundwater development is 200 per cent in Jhunjhunu, 197 per cent in Jodhpur, 181 per cent in Jalore, 168 per cent in Nagaur, 134 per cent in Sikar, 114 per cent in Pali and 104 per cent in Barmer (Central Ground Water Board, 2006). In the district of Pali, out of 10 assessed blocks, five have been categorized as over-exploited and five as critical. Similarly in district Nagaur, out of 11 assessed blocks, seven have been categorized as over-exploited, two as critical, one as semi-critical, and only one as safe. The average rate of depletion of groundwater table in the districts of Pali and Nagaur has been more than 0.4 metres per year. Further, increasing pollution of the water sources by the industrial units and over-abstraction of groundwater has led to water quality problems. Between 1996 and 2001, the number of villages and habitations suffering from quality problems in drinking water in arid western Rajasthan has gone up. The percentage was higher for habitations than for the village settlements, as the main villages are often covered by multiple or alternate sources (Rathore, 2007). In some districts, including Barmer, Bikaner, Churu, Ganganagar, and Jaisalmer, nearly all the habitations were affected by problems with the quality of the drinking water.

Socio-economic features of the region

The districts of western Rajasthan cover a total area of 2.09 lac km^2 and a population of around 2.25 crore persons. Population density (the number of persons per sq. km) is highest in Jhunjhunu (323) and lowest in Jaisalmer (13), which is the western-most district of the region. Joshi (2007) explained that the population density in Rajasthan is directly proportional to the average annual rainfall. Thus, as the rainfall decreases from east to west, the population density also displays significant variation. The total area under cultivation in the region is 109.20 lac ha, the highest being in Barmer (16.49 lac ha) and the lowest in Jhunjhunu (4.33 lac ha). The per capita cultivated land is highest in Jaisalmer (0.95 ha) and lowest in Sikar (0.23 ha). However, the land under cultivation as a percentage of the total area is highest in Ganganagar (71 per cent), which receives canal water for irrigation. The lowest percentage area under cultivation is in Jaisalmer (12 per cent), as much of the land area here is covered by sand dunes.

In the absence of any perennial surface water sources, more than 90 per cent of the water supply schemes in the region are based on groundwater. The net annual groundwater availability in western Rajasthan is 3,170.78 MCM, the highest being in Nagaur (548.38) and the lowest in Jaisalmer (60.09). The per capita groundwater availability is highest in Jalore (300 m^3/annum) and lowest in Churu (70 m^3/annum) (please refer to Table 6.1 for more details). The groundwater in the western region is already over-exploited, and there is high salinity and fluoride content, making it unfit for human consumption.

The Pali district covers a total area of 12,331 km^2, with a population of around 18.2 lac people, about 78.5 per cent of which live in rural areas. There are around 936 inhabited main villages in the district, most of which are covered with drinking water facilities (Census of India, 2001). The district has a total cropped area of 5,829 km^2, out of which only 1121 km^2 (19 per cent) is irrigated (Goyal et al., 2009). Nagaur (another district in which ACF is operating) covers a total area of 17,644 km^2, with a population of around 27.75 lac, about 83 per cent of which resides in rural areas. There are around 1,480 inhabited main villages in the district, of which 1,466 have drinking water facilities (Census of India, 2001). The district has a total cropped area of 13,654 km^2, out of which only 2948 km^2 (21.6 per cent) is irrigated (Goyal et al., 2009). The water supply schemes in both Pali and Nagaur districts are based on groundwater. However, the groundwater in these regions has excessive salinity and fluoride which pose a serious threat to public health. The fluoride content in the groundwater in the Nagaur district ranges from 3 ppm to 16 ppm. High levels of fluoride in groundwater that is used for human consumption can cause skeletal and dental fluorosis. The quality problem in drinking water

Table 6.1 Population, land use and groundwater availability in western Rajasthan

Districts	*Total area (km^2)*	*Population (000')*	*Population density (no./km^2)*	*Area under cultivation (000' ha)*	*Per capita cultivated land (ha)*	*Net annual groundwater availability (MCM)*	*Per capita groundwater availability (m^3/annum)*
Barmer	28387	1964.83	69	1649.22	0.84	256.45	130
Bikaner	27244	1674.27	61	1477.98	0.88	227.08	140
Churu	16830	1923.88	114	1151.22	0.60	128.98	70
Ganganagar	20634	3307.43	160	1476.32	0.45	312.52	90
Jaisalmer	38401	508.25	13	485.47	0.95	60.09	120
Jalore	10640	1448.94	136	648.62	0.48	432.33	300
Jhunjhunu	5928	1913.69	323	433.25	0.23	235.13	120
Jodhpur	22850	2886.50	126	1265.94	0.44	375.64	130
Nagaur	17718	2775.06	157	1237.48	0.44	548.38	200
Pali	12387	1820.25	147	576.22	0.32	282.16	150
Sikar	7732	2287.79	296	518.40	0.23	312.02	140
Total	208751	22510.89	108	10920.14	0.48	3170.78	140

Source: Data compiled from the Census of India, 2001; CGWB, 2006 and Joshi, 2007.

is more acute in the habitations of both districts, as compared to the main village settlements. Furthermore, the piped water supply is very erratic. Thus, the village community has a nominal dependence on the water supply schemes to meet its drinking water needs. Most of the domestic water requirements are met through village ponds and others sources, which include dug wells, tankers, etc.

The above discussion regarding the physical environment and socio-economic features clearly signifies the water scarcity problem existing in western Rajasthan. While the surface water resources are extremely limited, groundwater resources are also heavily over-exploited. Even this limited availability gets further reduced during the period of drought. The negative groundwater water balance indicates the problems this water-parched region will experience in the future. Under these extreme conditions, water conservation even at field- or micro-level can increase water availability at the local level. This chapter attempts to assess the hydrological and socio-economic impact of interventions carried out by the Ambuja Cement Foundation (ACF) near to its location at Jaitaran (in the Pali district) and Marwar Mundwa (in the Nagaur district).

Objectives and methodology

The main objective of this study is to evaluate the hydrological and socio-economic impact of reviving khadins and traditional village ponds. For the purpose of this study, a khadin constructed by ACF near Balada village (Jaitaran, district of Pali), two ACF-rehabilitated ponds near Balada and another near to the village of Marwar Mundwa (Mundwa, district of Nagaur) were selected. To understand the impact of these interventions, a random sampling of respondents, which included farmers, well owners and household occupants, was performed.

In respect of the khadin, five farmers cultivating their land inside the water spread area of the khadin were surveyed. For the village ponds, 30 households (HHs) dependent on the rehabilitated pond near Balada (the 'Balada pond') and 20 HHs dependent on a rehabilitated village pond near the village of Mundwa (the 'Lakholav pond') were selected. A random sampling technique was followed for the selection of respondents. Structured questionnaires addressing the various interventions were prepared and administered on the selected respondents in order to collect primary data. For secondary data, records from ACF officials and information from various reports were obtained and analyzed. A literature survey was performed, considering journal papers, scientific reports and data reports pertaining to the interventions. Group discussions were carried out in order to understand the existing situation of the region and resultant impact from the various interventions. Various statistical tools were used for the analysis of primary and secondary data.

The geo-hydrological impact of khadin and village ponds was analyzed by comparing the pre- and post-monsoon fluctuation in the water levels of those wells which were influenced by these interventions. The socio-economic impacts of khadin were assessed by analyzing a) the difference in input use, yield and net income from crops which benefited from khadin; and b) the difference in the area under crops, post-intervention. For village ponds, the socio-economic impacts were assessed by estimating a) the amount of time saved in fetching potable water from other common sources; b) the amount of money saved in the purchase of potable water from vendors; and c) the reduction in expenditure on medicines for mitigating the health impacts of exposure to poor quality water.

The work of ACF on khadins

The khadinis a traditional runoff farming system which is popular in western Rajasthan (Agarwal and Narain, 1997). This system has a great similarity to the irrigation methods that were practised in the Middle East, the Negev desert and south-western Colorado (Baindur, 2007; Prinz and Singh, undated). In khadins, the runoff from the upland surface is collected against an earthen embankment in the farmland which has a masonry waste weir for the outflow of runoff excess. The water-saturated khadin beds are then used for crop production. A rainfall of 75–100 mm is adequate to charge the khadin soils with sufficient soil moisture to raise a successful local crop (Prinz and Singh, undated). The soils in khadins are generally fertile due to the fine sediment deposited by the runoff.

A khadin farm is developed on the basis of rainfall probability, the available catchment area and its runoff generation potential. For efficient agriculture, a minimum 15:1 ratio of catchment area to crop area is required (Prinz and Singh, undated). The ponding of water in the khadin also induces groundwater recharge (Narain et al., 2005). The sub-surface water is extracted through bore wells developed inside the khadin or in the immediate vicinity downstream. A khadin of 20 ha in area, developed by CAZRI in the Baorali-Bambore watershed, has resulted in water conservation and provided an opportunity for farmers to grow kharif and rabi crops (Goyal, 2010). During the severe drought of 2002 in western Rajasthan, khadin farmers were able to meet domestic water needs and also grow sorghum for fodder. The average rise in the water level in shallow wells was 0.8 metres in sandstone and 2.2 metres in deep alluvium (Khan, 1996). Up to November 2010, ACF had constructed nearly eight khadinsnear to its Rabriyawas location and onekhadin near to its Marwar Mundwa location, with a total storage capacity of 2.13 million cubic metres. Recorded data from the observation wells (ten downstream and one upstream) under the influence zone of the Balada khadin, show an average rise of 2.31 metres in the water level between pre- and post-monsoon months (Figure 6.3).

The economic impact of the khadin

The Balada khadin was constructed in 2008 at a total cost of INR 3.5 lac. It basically consists of an embankment with a waste weir. The structure has a total catchment area of 1700 ha and an earthen embankment that is about 1.3 km long. The intervention has led to an increase in cropped area and well recharge benefits in the immediate vicinity. After the construction of the khadin, nearly 41 ha of land belonging to five households were brought under crop cultivation; the major crops

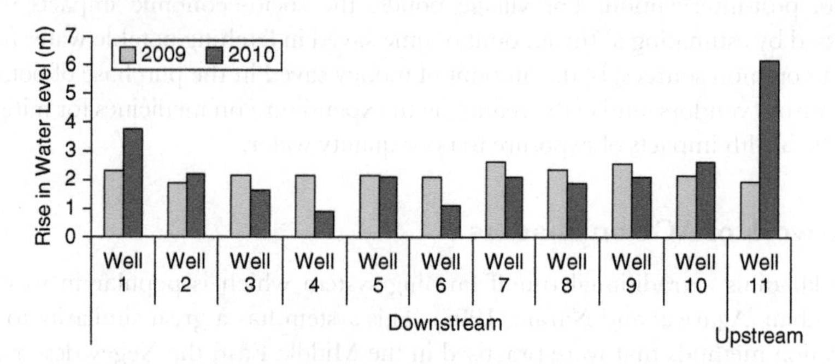

Figure 6.3 The rise in water level for observation wells under the influence of the khadin near Balada village.

Source: authors' own analysis using the secondary data.

in the water spread area included bajra, jowar and green gram. Presently, no irrigated crops are taken inside the khadin, but the farmers were confident of growing mustard and wheat in the coming years.

A comparison of mean yields for the crops (in kg/bigha, where 1 bigha equals 0.16 ha) shows that farmers were able to get 36 per cent higher yield for jowar and 25 per cent higher yield for green gram inside the khadin in comparison to the cropped land outside the khadin. Furthermore, the mean net returns (INR/bigha) (US $1 equates to INR 50) were higher for all the three crops taken inside the khadin. For bajra, it was nearly 9 per cent, for jowar it was 42 per cent and for green gram it was 54 per cent higher in comparison to the crops taken outside the khadin area (Table 6.2). The higher yields were attributed to a better soil moisture regime and the better availability of micro- and macro-nutrients in khadin land, which is endowed with nutrient-rich silt from its catchment. The higher yield increased the gross return, while the better nutrient content in the soil reduced the expenditure on fertilizers, and thus the input costs. The net result is a substantial improvement in net return. (Please note that the mean net incremental income figure provided in Table 6.2 is the difference between the mean net returns of the crops taken inside the khadin over that outside the khadin area.)

Table 6.2 Crop economics, inside and outside the water spread area of the khadin

Crops	*Bajra*		*Jowar*		*Green Gram*	
	Inside	*Outside*	*Inside*	*Outside*	*Inside*	*Outside*
Cropped area (bigha)	145	85	90	37	20	110
Irrigated area (bigha)	0	85	0	37	0	60
Mean input cost (INR/bigha)	2260	2890	2400	2808	1950	2850
Mean crop yield (kg/bigha)	440	500	410	300	250	200
Mean net return (INR/bigha)	7300	6710	7100	4991	8550	5550
Mean incremental income INR/bigha)	590		2109		3000	

Source: authors' own analysis using primary data.

From a benefit–cost point of view, the Balada khadin was found to be performing fairly well. The economic evaluation of the system assumed that the farmers would continue with the same cropping pattern and that there would be a normal monsoon for at least five out of the 15 years of life of the system (normal years are assumed to occur every three years). With these assumptions, the BC ratio (the direct benefit–cost ratio) for the khadin (for normal wet years) was estimated by taking the incremental net benefit from crop production using the khadin and the incremental cost of the system, including the capital and operating costs. The discount rate assumed for estimating the present worth of the costs and benefits was 7 per cent. The BC ratio comes out at 2.35 (Table 6.3). It is important to note that, although the expected active life of a khadin can be up to 15 years, the incremental income from the crops will vary significantly between the wet and dry years (when the crops yields get substantially reduced) in this water-scarce region. Here, we have assumed the benefits during the dry years to be zero.

The pond rehabilitation work of ACF

ACF started pond rehabilitation work near to its location at Rabriyawas (in the district of Pali) and Marwar Mundwa (in the district of Nagaur) in 2006–07. The main activities under the rehabilitation work included deepening the ponds and de-siltation. By 2010, ACF had rehabilitated 72 village ponds in Pali and 106 village ponds (including two mined-out areas) in Nagaur. The maximum number of ponds rehabilitated was in 2008–09. These works led to an average annual increase in the storage capacity of ponds by 0.40 million cubic metres. The capacity increase per village pond (in volume) was highest in 2010 (Figure 6.4). With respect to the availability of drinking water, around 79,230 people in Pali and 52,050 people in Nagaur districts benefited. Furthermore, a total of 1.65 lac livestock met their water needs in both the locations. Pond rehabilitation in Pali led to

Table 6.3 Benefit–cost ratio for the Balada khadin

Year	*Cost (INR)*			*Net incremental income (INR)*	*Present worth of the costs (INR)*	*Present worth of total Benefit in (INR)*
	Capital	*Repair and maintenance cost (INR) (10 per cent of the capital cost)*	*Total*			
0	3,50,000	—	3,50,000	—	3,50,000	—
1	—	35,000	35,000	3,35,360	32710.28	313420.56
4	—	35,000	35,000	3,35,360	26701.33	255844.54
7	—	35,000	35,000	3,35,360	21796.24	208845.35
10	—	35,000	35,000	3,35,360	17792.23	170480.02
13	—	35,000	35,000	3,35,360	14523.76	139162.48
Total			5,25,000	16,76,800	463523.00	1087752.95
BC Ratio					2.35	

Source: authors' own estimate using primary data.

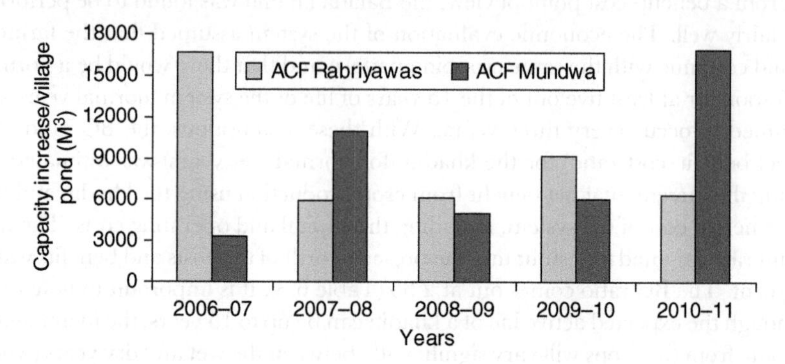

Figure 6.4 Capacity increase of village ponds (by volume), ACF's Rabriyawas and Mundwa locations.

Source: authors' own analysis using secondary data.

the recharge of 551 wells, benefiting 2,356 farmers. Through the recharged water, farmers were able to bring 2,820 more acres of land under cultivation. In Nagaur district, where silt from the ponds was used by the farmers, soil fertility improvement was achieved for a total of 290.45 acres of farm land.

The impact of pond rehabilitation

The Balada and Lakholav village ponds have a total storage capacity of 0.83 million cubic metres and 0.44 million cubic metres respectively. In the years of good rainfall, both ponds are fully filled with water. Traditionally, the village ponds are used for meeting year-round domestic and livestock water needs (given a normal wet year). Over the years, the capacity of these village ponds decreased, mainly because of poor or no maintenance and excessive siltation. The neglect of the ponds resulted in the village community receiving insufficient quantities of water from the ponds and no water at all during the peak summer months.

ACF started its rehabilitation work on the ponds in 2006–07. By 2010, the Balada village pond was deepened and de-silted five times and the Lakholav village pond was deepened and de-silted four times. The cost of rehabilitation was shared between ACF and the village community. This was necessary in order to get greater participation and make the village community responsible for the tank rehabilitation programme. The total cost of rehabilitation on Balada was estimated to be INR 22,84,800 (59.5 per cent contributed by villagers) and on Lakholav was INR 5,80,000 (62.5 per cent contributed by villagers). The unit cost of pond rehabilitation turned out to be INR 42/m^3 of additional storage created. This cost was shared by ACF and the village community in a 75:25 ratio. Contribution from the village community was mainly in the performance of manual labour work and the removal of silt from the ponds.

Condition of the pond after rehabilitation

As a result of the rehabilitation effort, the average annual increase in the storage capacity of Balada pond was 10,880 m^3 and that of Lakholav pond was 3,626 m^3. After rehabilitation, the ponds were able to meet the water needs of 11,500 people and a livestock population of 4,900. The benefit was greater in the case of the Lakholav pond, as this is the only source available to the village habitations of Marwar Mundwa for meeting their drinking water needs. Other sources in the village, which were mostly groundwater based, hada high fluoride content, thus making them unfit for human consumption.

Prior tore habilitation, water from the Balada pond was available only between the months of June and March, but, after rehabilitation, there was year-round availability of good quality drinking water. During the field visits, it was observed that the villagers were able to get water from the pond and from the wells recharged from the pond even during the month of May. Similarly, year-round availability of good quality water was noticed in the Lakholav pond. These results were further confirmed by 96 per cent of the respondents from Balada and 90 per cent of the respondents from Marwar Mundwa (for the Lakholav pond).

Hydrological impact of village pond rehabilitation

The rehabilitated Balada pond led to recharge of 27 wells in the surrounding area, benefiting 108 farmers. Through the recharged water, farmers were able to bring 120 more acres of land under cultivation. Recorded data from the 14 observation wells which were recharged by the Balada pond showed an average rise of 1.52 metres and 1.06 metres in the water level between the pre- and post-monsoon months of 2009 and 2010 respectively (Figure 6.5).

In 2009, the maximum rise in water level was observed in well no. 1 (2.74 metres) and the minimum in well no. 14 (0.3 metres). Whereas in the year 2010, the maximum rise in water level was observed in well no. 4 (1.9 m) and minimum in well no.12 (0.2 m). In the case of the rehabilitation of Lakholav pond, where silt

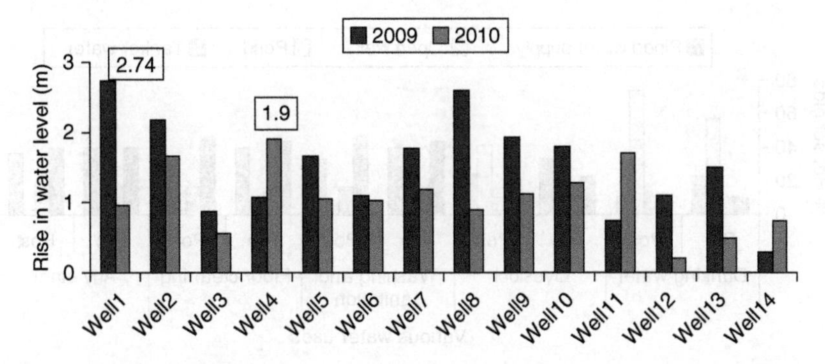

Figure 6.5 The rise in water level for observation wells recharged through Balada pond.

Source: authors' own analysis using the secondary data.

from the pond was used by the farmers, soil fertility improvement was achieved for a total of 2.3 acres of farm land.

Changes in water source and water use

A comparison of water usage from different sources pre- and post- rehabilitation shows that there was a shift in the preference of the source of water used for various purposes. It was observed that the number of households dependent on the pond near to Balada village to meet their drinking water needs increased from 17 to 22 (a 29 per cent increase) post-rehabilitation, whereas number of households which depended on the piped water supply had gone down from three to one (a 66 per cent decrease) and those dependent on tanker water had decreased from seven to three (a 57 per cent decrease). Also, the number of households collecting pond water for livestock purposes had increased from 10 to 12 (a 20 per cent increase) post-pond rehabilitation. However, no major change in the source of water for washing, sanitation, floor cleaning and other uses was observed. The shift in the choice of drinking water source towards the rehabilitated pond can be explained by the fact that other sources of water were highly contaminated. Moreover, only 40 per cent of the households had access to piped water supply, receiving water only once in every five days. No major change in water use for different activities was noticed but, certainly, the number of months for which the water is available from the ponds and from the wells recharged through ponds had increased (Figure 6.6).

At the pond located near to Marwar Mundwa village, no major changes in the sources of water were observed. Around 90 per cent of the households in the village habitation were totally dependent on the pond for all their water requirements, as there were high levels of fluoride in other sources of water. The piped water supply was highly unreliable, supplying water for about half an hour every alternate day. Villagers (as indicated by the response of all the sample households) did perceive that rehabilitation work had increased the storage capacity of the pond and hadalso led to an improvement in the quality of available water.

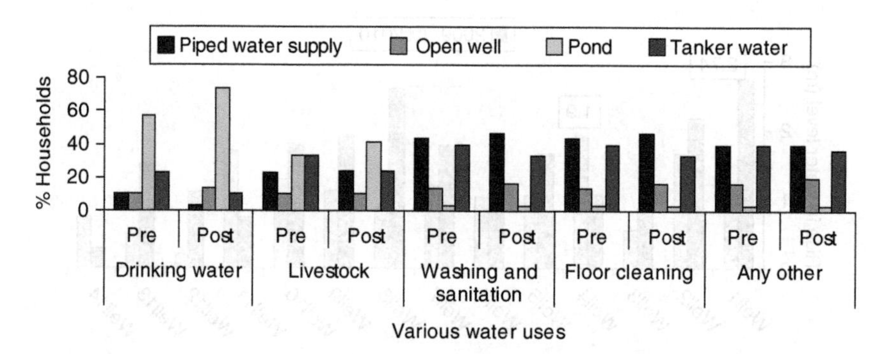

Figure 6.6 Change in water source for households located near Balada pond.

Source: authors' own analysis of primary data.

Overall, a significant change in water use was noticed post-rehabilitation (Table 6.4), in volumetric terms. The total water use per household had increased by 18 per cent, with the maximum increase in water use found in the case of water for drinking and cooking purposes (33 per cent), followed by livestock (21 per cent) and other uses which include washing, bathing and cleaning uses (11 per cent). These results clearly show the positive impact of pond rehabilitation on both the quality and quantity of the water available for domestic purposes, which had improved domestic water security and people's quality of life.

Household expenditure on water purchase and water-related diseases

Pond rehabilitation had a strong impact on the household expenditure on water purchase and the treatment of water-related diseases. For the village families located near the Balada pond, the annual expenditure per household on purchasing good quality water had reduced by around 14 per cent post-rehabilitation. This reduction was 67 per cent for the households located near the Lakholav pond. Furthermore, the annual expenditure on healthcare had reduced by 22 per cent post-Balada pond rehabilitation and 47 per cent post-Lakholav pond rehabilitation (Table 6.5). Prior to the ponds' rehabilitation, a major portion of healthcare expenditure was on the treatment of skeletal and dental fluorosis, which were

Table 6.4 Water use pattern for households located near Lakholav pond

Uses	*Water use (litres/household/day)*	
	Pre-rehabilitation	*Post-rehabilitation*
For drinking and cooking	66.25	88.25
For livestock	130.26	157.89
For other uses such as clothes washing, bathing, toilet use, floor and utensil cleaning	205.42	227.63
Total water use	401.93	473.78

Source: authors' own analysis using primary data.

Table 6.5 Expenditure pattern pre- and post-pond rehabilitation

	Balada Pond, Balada		*Lakholav Pond, Marwar Mundwa*	
	Pre-rehabilitation	*Post-rehabilitation*	*Pre-rehabilitation*	*Post-rehabilitation*
Annual expenditure on water purchase per household (INR)	2338.18	2014.29	1500	500
Annual expenditure on health per household (INR)	9158	7158	7500	4000
Total saving per household (INR)	2323.90		4500	

Source: authors' own analysis using primary data.

caused by the consumption of water having a high fluoride content. However, post-rehabilitation, a lower incidence of water-related diseases was reported by around 86 per cent of the households dependent on Balada pond and 100 per cent of the households dependent on Lakholav pond. These outcomes were attributed by the villagers to the availability of good quality water in larger quantities from the ponds after their rehabilitations.

Time spent in water collection

Pond rehabilitation work had actually helped the village community to drastically reduce the time spent on water collection. The households were able to save, on average, 3.4 hours per day post-Balada pond rehabilitation and 3.10 hours per day post-Lakholav pond rehabilitation. The maximum reported time saving was 5 hours and the minimum was 2 hours per day. A similar trend was seen in the case of RHWS in coastal Saurashtra, where the time saving for water collection was in the range of 1–7 hours. It should be inferred that the time saving has essentially come from reducing the distance travelled for fetching water from the piped water supply schemes, and the saving in the waiting time for water to be collected. Mostly, it is the women member of the households who go to fetch water from the village delivery point of the pipeline scheme. The time saving has actually increased the ability of parents to send their children school on time. This change was perceived by all the respondent households located near the rehabilitated Balada and Lakholav ponds.

Wage employment

The rehabilitation of the ponds also brought about significant changes in the household wage employment pattern. Post-rehabilitation of the Balada pond, the benefited households were able to find more time for doing various socio-economic production functions. Out of the 30 surveyed households, 24 (80 per cent) indicated increased time availability to work in their own farms, 27 (90 per cent) were able to give more time to household work and 10 (33.33 per cent) were able to find new wage labour opportunities. Similarly, in Lakholav, out of the 20 surveyed households, 13 (65 per cent) indicated increased time availability to work in their own farms, 20 (100 per cent) were able to give more time to household work and 15 (75 per cent) were able to find new wage labour opportunities. Table 6.6 presents seasonal data on the increased time available with the benefited households to undertaking various work. Increased availability of time to spend in farming is expected to result in increased agricultural productivity and increased time to work as wage labourers would lead to more income gains for the family.

The emergence of water sellers

Post-Balada pond rehabilitation, a few households in the neighborhood had even started selling water to other villages that were facing water shortage. These house-

Table 6.6 Seasonal increased time available, within the benefited households, to undertake various works

Type of work	*Balada village: mean increased time availability (hours per benefited household per day)*			*Marwar Mundwa village: mean increased time availability (hour per benefited household per day)*		
	Monsoon	*Winter*	*Summer*	*Monsoon*	*Winter*	*Summer*
For general household work	*Monsoon*	*Winter*	*Summer*	*Monsoon*	*Winter*	*Summer*
For work on own farm	1.67	1.71	1.67	1.46	1.46	1.46
For general household work	1.30	1.33	1.35	1.20	1.20	1.20
For wage labour	1.50	1.50	1.50	1.35	1.35	1.35

Source: authors' own analysis of the primary data.

holds were located very near to the pond and had wells which get recharged from the pond. One such household was found to be selling 15–20 tankers of water per day during peak summer months. The household has around six wells in the vicinity of the pond and sold water for a total of around 100 days per year. Each tanker contains 5,000 litres of water and was sold at the rate of INR 100 per tanker. There is a need to carefully monitor this development, as it may lead to inequity in water availability for the households within the village.

Major findings

Some of the important findings of this research study are:

- Wells under the influence zone of khadins and ponds were able to get the recharge benefits, indicated by a higher rise in water levels after monsoon.
- The irrigated area expansion was quite remarkable for khadin. Post-intervention, a greater proportion of the land was allocated to irrigated winter crops such as wheat, mustard and isabgol.
- There has been a substantial increase in the crop yield (kg/acre) and net income returns per unit of land (INR/acre) post-water interventions such as khadins. Also, the income returns (net) were better for the farmers located inside than for those located outside the influence zone of khadins.
- The benefit–cost ratio for one of the khadins was estimated to be 2.35. The estimates involved the assumption that the khadins would yield benefits only during wet years, which were assumed to be only five out of the total assumed khadin lifespan of 15 years.
- Post-rehabilitation, there was a year-round availability of good quality water (provided it was a normal or wet year) in the village ponds for the households, against 8 months pre-rehabilitation. The benefits were more significant for the habitations located near the Lakholav pond.
- The unit cost of pond rehabilitations is estimated to be INR 42 per cubic metre of additional storage created. A greater contribution from the village

communities ensured their participation and accountability in pond rehabilitation work.

- The groundwater recharge benefit of rehabilitation of Balada pond was remarkable. Twenty-seven wells got recharged, benefiting 108 farmers who were able to bring 120 more acres of land under cultivation. The average rise in water levels in wells around Balada pond were 1.52 metres and 1.06 metres between the pre- and post-monsoon months of 2009 and 2010 respectively.
- With pond rehabilitation, the dependence of village households on piped water schemes and tanker water supply for meeting their drinking and cooking needs had drastically come down. In Balada village, the number of households depending on the village pond to meet their drinking water needs increased by 29 per cent, whereas the number of households depending on a piped water supply and tanker water decreased by 66 per cent and 57 per cent respectively.
- Water use for domestic and livestock purposes increased, in volumetric terms, after rehabilitation. In Marwar Mundwa, the total water use per household increased by 18 per cent. The maximum increase was found for drinking and cooking use (33 per cent) followed by livestock (21 per cent) and other uses (11 per cent), which included washing, bathing and cleaning.
- The perceived physical, chemical and biological quality of water collected in the rehabilitated ponds was good. Improvement in quality of drinking water by the households resulted in reduction in expenditure on the purchase of good quality water and on healthcare, and lower incidence of water-related health problems. For instance, in Balada, the average annual expenditure on the purchase of good quality water and healthcare reduced by around 14 per cent and 22 per cent respectively.
- With pond rehabilitation, the households could reduce the time spent on water collection significantly: up to a minimum of two hours and a maximum of five hours in a day. The immediate outcome was that the adults from some of the families were able to take up additional wage employment, while members of the majority of the families were able to find more time to do farming operations. This would have significantly impacted family income.

Conclusions

The traditional water harvesting interventions undertaken by ACF in arid western Rajasthan has resulted in a beneficial impact on the region's groundwater balance as well as improvement in farmers' livelihoods and domestic water security. The experience of the farmers shows that the khadins have a great impact on agricultural production during wet years. They were able to bring more land under crop cultivation and received higher net returns. Farmers were also planning to sow-winter crops using the residual moisture present in the khadin bed. On the other hand, pond rehabilitation had improved domestic water security for the community in both the study villages. The design of the rehabilitation programme was such that it made village community responsible right at the initiation phase of the work. Not only were the impacts visible at the level of household, in terms of water security, but also with respect to the local groundwater regime and farm economy. With recharge of wells by rehabilitated ponds, the village communities were able to get drinking water even during severe summer months.

Improvements at the household water security level were manifested by a sharp fall, after pond rehabilitation, in the number of households depending on the long distance pipeline schemes and tanker water and by an increase in volumetric water use for domestic purposes. At the next level, with a significant reduction in the amount of time spent in water collection by the households, the families were able to send their children to school on time. With increased availability of time, the family members were also able to allocate more time for their farming operations. Additionally, consumption of good quality water from the ponds themselves meant that both household expenditure on the purchase of good quality water from other sources and medical expenditure reduced remarkably.

Strategizing small water harvesting in sub-Saharan Africa

Sub-Saharan Africa is another region in which rainwater harvesting in different forms, including runoff harvesting and roof-top rainwater collection tanks, is catching the imagination of international development organizations, development professionals and policy makers alike, as a local management initiative to improve the water security of poor rural communities for both crop protection and basic survival needs. But the economic viability of these systems has not been a part of the debate. As in the water-scarce regions of India, the hydrological regime of this vast geographical unit of Africa is characterized by a high degree of variability in rainfall and climate.

Rainfall variability, and the frequency of occurrence of droughts, is high in the drier regions of sub-Saharan Africa: Gommes and Petrassi (1994) showed that three out of the eight distinct regions of sub-Saharan Africa have a low rainfall index and experience high inter-annual variability and droughts. They are a) the Sahel and Sudan region, comprising Burkina Faso, Cape Verde, Chad, Gambia, Guinea-Bissau, Mali, Mauritania, Niger, Senegal and Sudan, which has high inter-annual variability; b) southern Africa, comprising Botswana, Lesotho, South Africa and Swaziland, has a relatively low rainfall index but also experiences the highest inter-annual variability in the whole of sub-Saharan Africa; and c) the Horn of Africa and Kenya region, comprising Djibouti, Ethiopia, Kenya and Somalia, is the driest region in sub-Saharan Africa, with low rainfall and high variability. They also found that the regions of Sahel and Sudan and the Horn of Africa and Kenya are also heavily drought-prone.

As seen in western Rajasthan, the hydrological extremes in these regions could affect the performance of rainwater harvesting systems, as they would cause excessively high inter-annual variability in runoff, with drastic reduction in stream flows during deficit years. Hence, during drought, the small water harvesting structures might fail to meet supplementary irrigation or domestic water needs, and may not, therefore, be dependable during critical years. The planning and economic evaluation of rainwater harvesting systems in these regions should take into account this limitation. The scope for integrating systems that are meant for meeting human and animal needs with large water storage systems needs to be explored in order to improve the dependability of the former. The cost–benefit analysis should consider typical wet, dry and normal rainfall years to capture the differences in economic returns due to the variation in hydrological benefits.

In the wetter regions, small water harvesting could be a good option for improving rural water security, but should be done with due consideration to catchment hydrology, topography and geo-hydrology. Gommes and Petrassi (1994) found that a) the southern-central Africa and Madagascar region, comprising Madagascar, Malawi, Mozambique, Namibia, Zambia and Zimbabwe, has comparatively less inter-annual variability in rainfall; b) the central Gulf of Guinea countries and Tanzania region, comprising Benin, Côte d'Ivoire, Ghana, Tanzania and Togo is not very drought-prone; c) the east and west Gulf of Guinea region, comprising Cameroon, Central African Republic, Equatorial Guinea, Gabon, Guinea, Liberia, Nigeria and Sierra Leone, is the wettest region in Africa with low rainfall variability; d) Angola, Congo and Zaire, the second wettest region in Africa, has a rainfall index of 1,489 mm and e) the great lake countries, comprising Burundi, Rwanda and Uganda, have high rainfall indices with low variability. Small water harvesting could be promoted in these regions, as a dependable source of rural water supplies for domestic and productive needs.

References

Agarwal, A., & Narain, S. (1997). *Dying wisdom: Rise and fall of traditional water harvesting systems.* New Delhi, India: Centre for Science and Environment.

Baindur, M. (2007). *When it rains on the sand dunes.* eSS working papers/knowledge studies. Bangalore, India: National Institute of Advanced Studies.

Central Ground Water Board (2006). *Dynamic ground water resources of India.* Faridabad, India: Ministry of Water Resources, Government of India.

Falkenmark, M., Lundquist, J., & Widstrand, C. (1989). Macro-scale water scarcity requires micro-scale approaches: Aspects of vulnerability in semi-arid development. *Natural Resources Forum, 13* (4), 258–267.

Goyal, R. K. (2010). Rainwater harvesting: A key to survival in hot arid zone of Rajasthan. In K. V. Rao, B. Venkateswarlu, K. L. Sahrawath, S. P. Wani, P. K. Mishra, S. Dixit, K. S. Reddy, M. Kumar, & U.S. Saikia (Eds.), *Proceedings of national workshop-cum-brain storming on rainwater harvesting and reuse through farm ponds: Experiences, issues and strategies* (pp. 29–38). Hyderabad, India: CRIDA.

Goyal, R. K, Angchok, D., Stobdan, T., Singh, S. B., & Kumar, H. (2009). Surface and groundwater resources of arid zone of India: Assessment and management. In A. Kar, B. K. Garg, M. P. Singh, & S. Kathju (Eds.), *Trends in arid zone research in India* (pp. 113–150). Jodhpur: Central Arid Zone Research Institute.

Gommes, R., & Petrassi, F. (1994). *Rainfall variability and drought in sub-Saharan Africa since 1960.* Agro meteorology working paper 9. Rome, Italy: Food and Agriculture Organization of the United Nations.

Joshi, K. N. (2007). Land use and land degradation in Rajasthan. In V. S. Vyas, S. Acharya, S. Singh and V. Sagar (Eds.), *Rajasthan: The quest for sustainable development* (pp. 77–100). New Delhi, India: Academic Foundation.

Khan, M. A. (1996). Inducement of groundwater recharge for sustainable development. In Indian Water Works Association (Ed.), *Proceedings of the 28th annual convention of the Indian Water Works Association* (pp. 147–150). Mumbai, India: IWWA.

Kumar, M. D. (2010). *Managing water in river basins: Hydrology, economics and institutions*. Delhi, India: Oxford University Press.

Narain, P., Khan, M. A., & Singh, G. (2005). *Potential for water conservation and harvesting against drought in Rajasthan, India*. Working paper 104, drought series paper 7. Colombo, Sri Lanka: International Water Management Institute.

Prinz, D., & Singh, A. K. (undated). *Technological potential for improvements of water harvesting*. South Africa: World Commission on Dams.

Rao, A. S. (2009). Climate variability and crop production in arid western Rajasthan. In A. Kar, B. K. Garg, M. P. Singh, & S. Kathju (Eds.), *Trends in arid zone research in India* (pp. 48–61). Jodhpur, India: Central Arid Zone Research Institute.

Rathore, M. S. (2007). Natural resource use: Environmental implications. In V. S. Vyas, S. Acharya, S. Singh, & V. Sagar (Eds.), *Rajasthan: The quest for sustainable development* (pp. 77–100). New Delhi, India: Academic Foundation.

Seckler, D. (1996). *The new era of water resources management: From "dry" to "wet" water savings*. Research report 1. Colombo, Sri Lanka: International Irrigation Management Institute.

Taylor, G. C., Roy, A. K., Sett, D. N., & Sen, B. N. (1955). Groundwater geology of the Pali region, Jodhpur division, western Rajasthan, India. *Bulletin of the Geological Society of India, B* (6) (engineering and groundwater).

7 The hydrological and farming system impacts of agricultural water management interventions in north Gujarat, India

O. P. Singh and M. Dinesh Kumar

Introduction

There have been numerous past research studies on the physical and socio-economic impacts of agricultural water management interventions. They broadly cover the following: the physical impact of water-saving technologies on irrigation water use (Narayanamoorthy, 2004); the impact of water-saving technologies and water-efficient crops on crop water productivity in physical terms (kg/m^3) (Kumar, 2007); the benefit–cost analysis of micro-irrigation systems, such as drips and sprinklers (Palanisamy et al., 2002; Kumar et al., 2004; Narayanamoorthy, 2004) and the comparative economics of the cultivation of water-efficient and high valued crops; and only limited analysis of the economic and social costs and benefits of micro-irrigation systems (Suresh Kumar and Palanisami, 2011). But all these analyses are based on individual plot level assessments of physical, economic, environmental or social variables.

However, the introduction of micro-irrigation systems or agricultural water management technologies can change the dynamic of the entire farming system (Kumar et al., 2008; Kumar, 2009). For instance, the adoption of a micro-irrigation system is associated with farmers shifting to crops that are amenable to these systems. This means that the water-saving impact will be the sum total of the potential improvement in efficiency of the use of water for a particular crop resulting from technology adoption, but also from the change in crop water requirement (ET) itself owing to a change in crop in the aftermath of technology adoption.

Often, the adoption of high valued crops is associated with the introduction of skilled labour hired from outside, which replaces domestic labour; and the mechanization of farms (Dhawan, 2000). If the adopter family is not able to divert the saved domestic labour to other production functions, system adoption can actually lead to an increase in input costs instead of the saving in labour cost that is often projected as a benefit of MI systems. Shifts in cropping pattern can potentially impact on the livestock holding of farmers, milk production and income from dairying, and the overall composition of farm economy (Kumar, 2007; Kumar and Amarasinghe, 2009). Hence, individual plot level assessments of physical and socio-economic impacts can often be misleading.

There is a need to understand the overall changes in the farming system resulting from adoption of MI systems and high valued crops. A related concern is how use of groundwater for agriculture in a region changes as a result of the adoption of water-saving irrigation technologies and water-efficient crops. While pursuing the goal of sustainable groundwater management, the fact that groundwater depletion affects the poor farmers more adversely (Dubash, 2000; Kumar, 2007) means such concerns are valid.

The North Gujarat Sustainable Groundwater Management Initiative

Groundwater over-exploitation is a phenomenon found in many arid and semi-arid regions of the world (Custodio, 2000). With an annual draft of 231 BCM, India has the highest groundwater withdrawal for agriculture (Kumar, 2007). If one goes by the official estimates of groundwater development, which considers only the hydrological data, only 23.1 million hectare metres out of the 43.2 million hectare metres of renewable groundwater in the country is currently utilized. But, if one goes by the disaggregated data, only 15 per cent (839) of the blocks/talukas/mandals in the country are over-exploited; 4 per cent are critically exploited and 10 per cent (550) are in the semi-critical stage (GOI, 2005).

Within India, north Gujarat is one of the intensively exploited regions. Groundwater supports irrigated crop production and intensive dairy farming in the region. Well irrigation is critical to the region's rural economy and livelihoods (Kumar, 2007; Singh et al., 2004). Hence, the management of groundwater is crucial for the survival of the rural communities in that region.

Internationally, discussions on approaches to managing groundwater/aquifers include enforcement of tradable property rights (Rosegrant and Schleyer, 1996; Thobani, 1997); metering agricultural pumpsets and energy pricing (Kumar, 2007; Saleth, 1997; Zekri, 2008); creating local management regimes with a nested hierarchy of institutions from village to aquifers, along with tradable water rights (Kumar, 2000, 2007); energy rationing (Zekri, 2008); decentralized water harvesting and recharge (Shah et al., 2003); and conjunctive management of groundwater using water from large surface reservoirs (Contor, 2009; Llamas, 2000; and Ranade and Kumar, 2004).

IWMI's initiative in north Gujarat, which started in 2002 under the IWMI-Tata Water Policy Research Program, explored farmer-initiated agricultural water demand management as a strategy to reduce the stress on groundwater resource in the region. The beginning of the strategy was improving water productivity in agriculture. The North Gujarat Initiative, currently being managed by SOFILWM (Society for Integrated Land and Water Management), focused on introducing water-efficient irrigation technologies, crops that give high returns per unit of both land and water, and practices that improve the primary productivity of land.

Groundwater management strategy in north Gujarat

According to some estimates done for the White Paper on Water in Gujarat, the total water used in agriculture in the region was 5,372.5 MCM in 1996–97 (IRMA/UNICEF, 2001). On the other hand, while the total renewable water resources of the region is 6,105 MCM (as per GOG, 1996, cited in IRMA/UNICEF, 2001), the total water use was estimated to be 6,008 MCM as far back as 1996–97 (Table 8: IRMA/UNICEF, 2001). The per capita annual water withdrawals in north Gujarat exceeded the renewable water availability by the year 2000 (Kumar and Singh, 2001).

From these figures and the earlier statement that the basins in north Gujarat are closed, it is clear that supply-side approaches to deal with groundwater depletion problems are not going to make any impact on the region's groundwater regime, and the only solution lies in water demand management. Since agriculture takes the lion's share of the total water diverted from surface systems and aquifers in the region (nearly 92 per cent), water demand management in agriculture was chosen as the strategy for improving the demand–supply balance in groundwater of the region. This was done to achieve long-term sustainability through enhancing water productivity in the sector. Three specific interventions were identified to achieve water productivity improvement: a) use of efficient irrigation technologies for crops, which helps improve the crop yields and reduce the consumptive water use (depleted water); b) the introduction of crop that are highly water-efficient in terms of net return per unit of water consumed (INR/ET); and c) improvement of the primary productivity of land through improvement soil nutrient management measures.

The theoretical foundations for the strategy were twofold. First, the use of micro-irrigation devices would reduce the actual amount of water depleted in crop production. While this is in accordance with conventional wisdom, internationally, the concept of using micro-irrigation to reduce consumptive use of water for crop production and water saving in agriculture has not been widely recognized. On the contrary, as argued by Molle and Turral (2004), some scholars believe that use of micro-irrigation systems would eventually increase the consumptive use of water. Their contention is that, while the amount of water applied to crops could be reduced through efficiency improvements, the consumptive water use remains the same: since the farmers perceive a reduction in the amount of water pumped for irrigation, they might expand the area under irrigation, there by increasing the consumptive use of water.

Table 7.1 Per capita renewable freshwater availability in Gujarat, by region

Name of the region	*Total freshwater availability (MCM)*
South and central Gujarat	37926
North Gujarat	6105
Saurashtra	9287
Kachchh	1275
Gujarat	54593

Source: IRMA/UNICEF, 2001 White Paper on Water in Gujarat.

But, in the case of north Gujarat, the hydrology of water use is different. The water, which goes into deep percolation under conventional methods of irrigation, is "non-recoverable" as eventually part of it gets lost in non-beneficial soil evaporation (after the land becomes fallow) and the remaining part gets held up in the unsaturated zone as hygroscopic water (see Allen et al., 1998, for details). In sum, use of micro-irrigation technologies would reduce the consumptive fraction (CF), leading to real water saving (Kumar et al., 2008). Furthermore, the issue of return flow is less relevant for row crops, in which the non-beneficial soil evaporation from the land which is not covered by the crop canopy can be reduced using technologies like drip irrigation. Hence, the real water saving would be greater in the case of drip systems used for row crops (Kumar et al., 2008). Second, the use of water-efficient crops that give higher returns per unit of land and water would also help towards reducing the depletion of groundwater.

Major achievements

The physical achievements made in the region, as a result of these various interventions over the past seven years, are summarized in Table 7.2. The total area under sprinkler and drip irrigation systems in the villages selected for field interventions and in the villages falling outside the project area was around 24,285 ha.

Objectives and scope of the research

The objectives of the research were:

- To study the water demand management interventions being adopted by different categories of farmers, such as small/marginal, medium and large farmers in the north Gujarat region.
- To analyze the impact of these interventions on the farming system, livelihood patterns, food and nutritional security, poverty and gender division of labour for different categories of farm households.

Table 7.2 Key physical achievements of the North Gujarat Initiative

Sr. no.	*Type of activity*	*No. of farmers in the project villages*	*Total area in the project villages (ha)*	*Total no. of farmers outside the project area*	*Total area outside the project villages (ha)*
1	Drip irrigation	656	1519.0		
2	Sprinkler irrigation	542	1229.0	10,689	21,537.0
3	Plastic mulching	15	62.1		N/A
4	Organic farming	801	792.0		N/A
5	Horticulture	680	320.0		N/A
6	Drum kit	411	411.0		
7	Vegetable kits	1670	1670.0		N/A

Source: SOFILWM office records.

- To analyze the potential impact of the combination of water demand management interventions for different scales of implementation on agricultural surpluses and groundwater use, and assess their implications for food security, risk and the vulnerability of farming communities, and labour absorption.

The study covered 49 villages of eight talukas from two districts of north Gujarat (Banaskantha and Mehsana), covering a total of 114 adopter farmers and 51 non-adopter farmers. The sample farmers were picked from the alluvial areas in the semi-arid and arid parts of the region, and, hence, the findings are more relevant for such areas. The study analyzed the impacts of various interventions at the plot, field and farm level.

Approach, methods and tools

The approach used in the study involved comparing the plots, fields and farms of farmers before and after the adoption of new crops and water-saving irrigation technologies. The variables considered for comparison were the overall cropping pattern, gross cropped area of the farm and area under different crops; livestock composition and size; the water application rate for individual plots of crops; the level of crop inputs and the cost; yield and net return from different crops; and the inputs and outputs for different types of livestock.

The BC ratio for micro-irrigated (MI) crop is *i* calculated as:

$$\left\{NI_{MI\text{-}irrigation,i} - NI_{trad\text{-}irrigation,i}\right\} \Big/ C_{MI,i} \quad (1)$$

Here, *NI* is the net income from one hectare of the crop grown in the plot, and suffixes *MI-irrigation* and *trad-irrigation* stand for crops irrigated by MI system and crops under traditional method of irrigation respectively. $C_{MI,i}$ is the annualized capital cost of the system for one hectare, apportioned among all the crops grown during the year with the same system. Obviously, if two crops are grown during the same year (for instance, groundnut in summer followed by potato in winter), the annualized cost of the MI system was apportioned among them.

The net income from the crop *i* (NI_i) is calculated as:

$$NI_i = GI_i - IC_i \quad (2)$$

Here, GI_i and IC_i stand for gross income and input costs per hectare of the crop, respectively for crop *i*. Nevertheless, while estimating the input costs, the capital cost of the MI system should not be considered. The same is taken into account for estimating the modified net income, and is estimated as:

$$NI^1_{i_i} = NI_i - C_{MI,i} \quad (3)$$

The total farm level water saving WS_{FARM} (cubic metres) owing to the adoption of MI systems and water-efficient crops is estimated as:

$$WS_{FARM} = 10000 * \{\Sigma_{i=1}^{m} A_i * \Delta_i - \Sigma_{j=1}^{n} A_j * \Delta_j\} \quad (4)$$

Here, A_i stand for the area under crop *i* in hectares, grown in the farm in the pre-MI-adoption case; A_j stand for area under crop *j* grown in the post-MI adoption phase. The suffixes m and n stand for the number of crops grown in the pre-adoption phase and post-MI adoption phase, respectively. Δ_i and Δ_j are the irrigation water applied for crops *i* and *j*, respectively, in cubic metres per ha. The area figures are averages estimated for the entire sample of 114 farmers. Hence, the water saving estimated would be for an average farm.

The physical productivity of water in crop production θ_i (kg per cubic metre) for crop *i* was estimated as:

$$\theta_i = Y_i / \Delta_i \quad (5)$$

Here, Y_i is the yield of crop *i* (kg per hectare); and Δ_i is as explained above.

The economic productivity of water in crop production ∂_i (INR per cubic metre) for crop *i* was estimated as:

$$\partial_i = NI_i / \Delta_i \quad (6)$$

While estimating the economic productivity of water for crops irrigated by MI, the modified net income was considered (see Equation (3)).

The regional level water saving) ($WS_{REGIONAL}$) through a combination of agricultural water management interventions is estimated by multiplying the average water saving per individual farm (WS_{FARM}) by the total number of farms under micro-irrigation. The second variable is estimated on the basis of the total area under MI systems in the north Gujarat region, and the average size of the MI-irrigated plot in the sample farm. Using such a methodology, the error in estimation would be high if the sample farms are not representative of the regional situation, in terms of the proportion of the total farm under MI systems. The underlying assumption in the estimation is that all the water applied to the crop iseventually depleted from the system. This is a reasonable assumption given the semi-arid to arid climate and deep unsaturated zone in the region's aquifers.

$$WS_{REGIONAL} = WS_{FARM} * TAREA_{MI} / AREA_{MI,FARM} \quad (7)$$

Here, $TAREA_{MI}$ and $AREA_{MI,FARM}$ are the area under MI system in north Gujarat region and average area under MI system in the sample farm, respectively.

The change in the overall net return from farming can be estimated as:

$$NI_{FARM} = \{\Sigma_{j=1}^{n} A_j * NI_j + \Sigma_{l=1}^{p} N_p * LI_p - \Sigma_{i=1}^{m} A_i * NI_i - \Sigma_{k=1}^{o} N_k * LI_k\} \quad (8)$$

Here, A_i is the area under crop *i* which is not MI irrigated; A_j is the area under crop *j* which is MI-irrigated; m and n are the number of crops grown by farmers before adoption and after adoption, respectively. N_k and LI_k stand for the total number of livestock belonging to the category k, and the net income per annum from one animal belonging to that category, respectively. The suffixes o and p stand for the total number of livestock categories owned by the farmers before and after the adoption of the MI system.

Data sources and types

The major source of data was the primary survey of adopter and non-adopter farmers in the north Gujarat region. The types of data included the inputs and outputs of all the crops grown and different types of livestock reared by the adopter farmers, including those which are not covered by MI systems; and the inputs and outputs of all the crops grown and livestock reared by the non-adopters. The data for adopters included that prior to adoption as well. The data are: a) the area under each crop; b) the inputs such as seed cost and labour (days); c) the cost of fertilizer and pesticide used; d) the number of watering and hours of irrigations for each watering (hours per irrigation per ha); e) the number of different types of livestock, and average feed and fodder (both dry and green) inputs for various types of livestock (per animal per day; f) the yield of various crops, including both main product and by-product (kg/ha); g) the average milk outputs for different animals (litre per day); and h) the cost of various MI systems (INR per hectare).

Results and discussions

Who are the adopters?

The average family size and farm holding size of the adopter and non-adopter farmers is given in Table 7.3 and Table 7.4, respectively.

Table 7.3 Average family size of adopters and non-adopters of water-saving technology

Particulars	*Total family size*	*Adult male*	*Adult female*	*Children*	
				Male	*Female*
Adopter	8.22	2.58	2.61	1.68	1.34
Non-adopter	6.83	2.40	2.48	1.08	0.90

Source: authors' own estimates based on primary data.

Table 7.4 Average farm holdings of adopters and non-adopters of water-saving technology

Particulars	*Total land holding size*	*Cultivable land*	*Cultivated land*	*Irrigated land*
Adopter	3.79	3.76	3.74	3.74
Non-adopter	2.76	2.75	2.74	2.68

Source: authors' own estimates based on primary data.

Changes in individual components of the farming system

The individual components of the farming system considered for analysis are the cropping pattern; the crop yields; the different types of water-efficient irrigation systems and their capital costs; the irrigation intensity with and without MI system; the area under forage crops; the area under orchards; the livestock holding; and the gross and net outputs from crops and gross return from dairying. They are analyzed separately in the subsequent sections for changes in irrigation water use; changes in crop yield; changes in cropping pattern and livestock composition; changes in net return from the entire farm with structural changes, as well as for individual crops and livestock categories; and a BC analysis of different MI technologies, and farming system level changes in irrigation water use, is undertaken.

Changes in water application for different crops

As noted by Kumar et al. (2008), the real water saving through the use of micro-irrigation systems is a function of the crop grown, the soil type, the type of MI technology, and the climate and geo-hydrology. Therefore, applied water saving would also be a function of the first three factors. In situations like north Gujarat, the most perceptible impact of adoption of a MI system is likely to be applied water saving, as it would be high in semi-arid and arid climate, sandy soils, and for row crops. The saving would be more for drip-irrigated row crops due to the reduction in non-beneficial soil evaporation (based on Allen et al., 1998; Kumar et al., 2008).

Table 7.5 shows that, with the adoption of a MI system, the total irrigation water application rate had reduced significantly for most of the crops. The

Table 7.5 Irrigation water use for different crops before and after adoption of MI

Season	*Crop*	*Method of irrigation*	*Irrigation water use (M^3/Ha)*
Before adoption of water-saving technology (WST)			
Monsoon	Cluster bean		2549.00
	Castor		7890.10
	Groundnut		5602.80
	Chilli		11500.00
	Brinjal		5966.70
	Green gram		840.00
	Cotton		7150.60
	Fennel	Traditional method of irrigation	2455.25
Winter	Mustard		6337.01
	Potato		13964.90
	Rajgaro		3600.00
Summer	Pearl millet		8368.20
	Millet		11338.60
	Fodder bajra		20850.00
	Vegetable		13750.00

After adoption of water-saving technology			
Monsoon	Cluster bean	Sprinkler	1305.00
	Castor	Drip	7695.00
	Groundnut	Sprinkler	5258.20
	Chilli	Drip	3540.00
	Alfalfa	Sprinkler	12815.10
	Brinjal	Drip	1180.00
	Kola	Drip	540.00
	Pomegranate	Drip	3334.0
	Cotton	Drip	3510.00
	Fennel	Drip	1728.00
Winter	Tomato	Drip	9440.00
	Potato	Sprinkler	12721.40
	Flower	Sprinkler	3540.00
Summer	Pearl millet	Drip	5030.80
	Millet	Sprinkler	8776.10
	Choli	Sprinkler	5611.50

Source: authors' own estimates based on primary data.

reduction is more than 50 per cent in some cases, while insignificant in some others. As we have already pointed out, the extent of the reduction is a function of the technology used for irrigation. This, again, is determined by the crop. For most vegetables, drip irrigation is used (chilli, tomato and brinjal). For potato, cluster bean and groundnut, micro sprinklers are used. For cotton and castor, drip irrigation is used. For bajra and cluster beans, overhead and mini sprinklers are used.

As regards actual impact, in the case of cluster beans the water application rate dropped from 254.8 mm to 130.5 mm. In the case of cotton, the extent of reduction is more than 50 per cent from 715 mm to 351 mm. In the case of chilli, the extent of the reduction was nearly 70 per cent (i.e., from 1150 mm to 354 mm). This is an exceptionally high value. In the case of summer bajra (pearl millet), the water application rate reduced from 836.8 mm to 503 mm. The total water application rate for pomegranate was estimated to be 333 mm, but, this is a crop introduced with the MI system, and data on the irrigation water use rate without a MI system are not available.

For potato, the water application rate was found to be excessively high when compared to the fact that it is a short duration crop (90–100 days) of winter. The main reasons for this could be that the area in which the crop is predominantly grown has very light sandy soils with a high rate of soil infiltration. This means that a substantial amount of water is lost in deep percolation even under the sprinkler method of irrigation.

Changes in the yield of crops sold

Analyses of crop yields show some interesting trends. For most crops – cluster bean, castor, chilly, cotton, fennel, wheat and groundnut – the yield was higher under the MI system. In the case of castor and fennel, the increase in yield was more than 50 per cent. In the case of chilli, the yield increase was 25 per cent. But, for some crops such as brinjal and summer bajra, the yield was lower under micro-irrigation. In the case of summer bajra, this phenomenon of reduced yield with micro-irrigation can be explained by the poor distributional uniformity obtained in water application through the overhead sprinklers.

But these unusual findings with regard to yield in no way mean that, with the adoption of MI systems, the yield for these crops can go down: the figures presented in Table 7.6 are averages for those who grew the crops with MI systems and those who grew without, and the farmers who showed lower yields under MI systems are not necessarily the same as those who showed a higher yield without MI, although there may be some farmers in common. The results lead us to the importance of agronomic practices – such as use of nitrogenous fertilizers and the provision of adequate irrigation to meet the crop water requirement – in obtaining higher yields, as well as the use of MI systems.

Changes in the area under different crops

Some remarkable changes in the area under crops were noticed after the adoption of MI systems. The average gross cropped area reduced from 4.07 ha to

Table 7.6 Yield of irrigated crops with and without MI systems

Before adoption of WST			*After adoption of WST*		
Season	*Crop*	*Average yield (Qt/Ha)*	*Season*	*Crop*	*Average yield (Qt/Ha)*
Kharif	Cluster bean	14.34	Kharif	Cluster bean	15.00
	Castor	21.40		Castor	33.33
	Groundnut	20.80		Groundnut	21.78
	Chilli	600.00		Chilli	750.00
	Alfalfa	NA		Alfalfa	1620.00
	Brinjal	466.67		Brinjal	250.00
	Cotton	32.72		Cotton	39.71
	Fennel	7.17		Fennel	15.84
	Bajra	16.67		Kola	60.00
	Green gram	12.00		Pomegranate	42.03
Winter	Wheat	37.98	Winter	Wheat	50.00
	Potato	337.37		Potato	345.34
	Rajgaro	4.00		Flower	100.00
	Mustard	32.43		Tomato	1200.00
Summer	Bajra	48.97	Summer	Bajra	40.68
	Millet (Jowar)	59.00		Millet (Jowar)	55.18
	Vegetable	50.00		Chickpea	39.93
	Fodder bajra	875.00		Groundnut	45.00
	Groundnut	25.00			

Source: authors' own estimates based on primary data.

3.21 ha, but these changes seem to affect only select crops. Table 7.7 shows that both the absolute and percentage area under potato, kharif groundnut, vegetables and alfalfa had increased significantly. Also, tomato appears as a winter crop in the post-adoption scenario. But, on the other hand, both the absolute area and percentage under bajra and wheat reduced substantially, while the area under mustard became nil. The reduction in the area under wheat, millet, pearl millet and rajgaro was quite remarkable. These are crops which are grown mainly for domestic consumption, as wheat and bajra are part of the staple food. (In particular, rajgaro cooked in milk is used for feeding children.) Hence, a reduction in their area will have some implications for domestic food security in the immediate term, given that the prices of cereals shot up during 2009, the year prior to the study.

Potato, groundnut and chilli are amenable to micro-irrigation systems, and farmers in the area are extensively irrigating these crops with MI systems. This observation validates our assumption that, after realizing the benefits of the adoption of MI systems, farmers tend to allocate a greater area of their farms to those

Table 7.7 The area under different crops of adopters before and after adoption of MI

Kharif		*Winter*		*Summer*	
Crop	*Area (Ha)*	*Crop*	*Area (Ha)*	*Crop*	*Area (Ha)*
Before adoption of WST					
1. Cotton	0.118	1. Potato	1.016	1. Millet (Jowar)	0.146
2. Castor	0.288	2. Wheat	0.148	2. Pearl millet	0.921
3. Fennel	0.028	3. Rajgaro	0.018	3. Vegetable	0.005
4. Groundnut	0.835	4. Mustard	0.177	4. Groundnut	0.007
5. Chilli	0.004			5. Fodder bajra	0.019
6. Brinjal	0.013				
7. Alfalfa	0.117				
8. Cluster bean	0.174				
9. Sesamum	0.026				
10. Pearl millet	0.011				
11. Green gram	0.004				
Gross cropped area					**4.074**
After adoption of WST					
1. Cotton	0.106	1. Potato	1.469	1. Millet (Jowar)	0.031
2. Castor	0.014	2. Wheat	0.019	2. Pearl millet	0.172
3. Fennel	0.020	3. Flower	0.004	3. Vegetable	0.039
4. Pomegranate	0.047	4. Tomato	0.011	4. Groundnut	0.007
5. Groundnut	1.109				
6. Chilli	0.011				
7. Brinjal	0.003				
8. Alfalfa	0.122				
9. Cluster bean	0.009				
10. Sesamum	0.014				
11. Kola	0.004				
Gross cropped area					3.211

Source: authors' own estimates based on primary data.

crops that are amenable to MI systems and for which they obtained good results with MI systems.

Changes in the inputs and outputs of livestock

Table 7.8 shows that the average number of milk animals (per adopter farmer) belonging to all the three categories of livestock (buffalo, cross-bred cow and indigenous cow), had increased post-adoption, though the rise was not substantial. More importantly, the average milk yield has also gone up for all of the three categories of livestock, with a significant increase in the case of cross-bred cows. The price of milk had also gone up over the years. Hence, the gross income from milk production had gone up significantly. But what is important from the point of view of our analysis is the differential income due to the increase in milk output per animal and the increase in holding size, rather than the rise in price. This may be attributed to the increased availability of green fodder (from alfalfa and other forage crops grown by the farmers), caused by both an expansion of the area under those crops and, as a result of the MI system adoption, increased crops yields.

A close look at the fodder cultivation practices of the adopters and non-adopters illustrates this. In spite of fewer farmers growing alfalfa after adoption, the average area per family (worked out on the basis of the total number of adopters, i.e., 114) is still higher (0.122 ha against 0.117 ha). Also, around 18 farmers were using sprinkler and drip for the crop, and 15 were using sprinklers for fodder bajra. Earlier studies have shown the yield impact of micro-irrigation systems on alfalfa in the region (Kumar et al., 2008a). This, also, might have contributed to the increasing availability of green fodder in the adopter households at the farm level.

Changes in net return and the water productivity of different crops

Table 7.9 provides the mean values of net income, modified net return and the water productivity of the crops without MI systems and with MI systems. The

Table 7.8 Yield and gross income obtained by farmers from different types of livestock before and after adoption of MI systems

Type of animal	*Total no. of in-milk animals*	*Total milk production (Lt/day)*	*Milk price (INR/Lt)*	*Dry animals*	*Calves*	*Gross income (INR/day)*
Before adoption of WST						
1. Buffalo	2.29	17.06	14.80	1.05	1.65	252.44
2. Cross-bred cow	0.84	8.27	11.05	0.27	0.52	91.43
3. Indigenous cow	0.08	0.61	10.00	0.01	0.06	6.05
After adoption of WST						
1. Buffalo	2.38	17.47	18.41	1.07	2.05	321.66
2. Cross-bredcow	1.04	11.86	12.04	0.25	0.81	142.77
3. Indigenous cow	0.09	0.74	10.80	0.08	0.09	7.96

Source: authors' own estimates based on primary data.

Table 7.9 Net income, modified net income and water productivity, in physical and economic terms, with and without the adoption of MI systems

Season	*Crop*	*Type of technology used for irrigation*	*Net return (INR/Ha)*	*Modified net return (INR/Ha)*	*Water productivity Physical (Kg/m³)*	*Water productivity Economic NR/m³)*
Before adoption of WST						
Monsoon	Cluster bean		13194.24	13194.24	0.56	7.68
	Castor		21070.10	21070.10	0.27	3.04
	Groundnut		11133.74	11133.74	0.37	4.13
	Chilli		411833.33	411833.33	5.22	34.90
	Brinjal		157533.33	157533.33	7.82	44.91
	Pearl millet		4663.33	4663.33	0.13	0.76
	Green gram		4450.00	4450.00	1.43	5.30
	Cotton	Traditional method of irrigation	68876.42	68876.42	0.46	10.30
	Fennel		12333.33	12333.33	0.29	6.30
Winter	Mustard		43994.00	43994.00	0.51	8.00
	Wheat		23195.36	23195.36	0.47	4.58
	Potato		60684.85	60684.85	2.42	7.04
	Rajgaro		4182.00	4182.00	0.11	1.16
Summer	Pearl Millet		19771.10	19771.10	0.27	3.49
	Millet		26797.62	26797.62	0.52	2.15
	Fodder bajra		28583.33	28583.33	4.20	1.56
	Vegetable		16166.67	16166.67	0.36	1.18
After adoption of WST						
Monsoon	Cluster bean	Sprinkler	20575.00	17811.55	1.15	13.65
	Castor	Drip	51150.00	40360.51	0.43	5.43
	Groundnut	Sprinkler	27894.17	24039.10	0.41	7.70
	Chilli	Drip	524250.00	520162.19	21.20	146.90
	Alfalfa	Sprinkler	55349.57	48513.63	12.60	5.67
	Brinjal	Drip	86650.00	82562.19	21.20	119.00
	Kola	Drip	9800.00	6559.74	7.41	12.15
	Pomegranate	Drip	81662.50	67988.34	1.26	37.80
	Cotton	Drip	52822.88	29617.54	1.13	12.44
	Fennel	Drip	23730.29	18034.76	0.92	36.91
Winter	Tomato	Drip	475000.00	469646.10	12.70	49.75
	Wheat	Sprinkler	53361.11	51273.13	1.70	26.19
	Potato	Sprinkler	98024.13	93538.60	3.10	11.39
	Flower	Drip	5000.00	1430.74	2.80	0.40
Summer	Pearl millet	Sprinkler	15082.45	12494.82	0.81	3.84
	Millet	Sprinkler	22099.55	19458.53	0.63	2.66
	Choli	Drip	22564.00	17279.54	0.71	12.94
	Groundnut	Sprinkler	86250.00	83289.00	0.38	7.09

Source: authors' own estimates based on primary data.

modified net returns are obtained by subtracting the annualized cost of the micro-irrigation system from the net return for the crops. Therefore, for pre-adoption conditions, it is same as the net return. As expected, the average net returns are higher under MI systems for all the crops except brinjal and cotton. We have already seen that, in the case of brinjal, the average yield of this crop for irrigated plots was slightly lower. This might have resulted in lower net income. In the case of cotton, although the yield was higher under the MI system, the net income was lower. This was due to the higher input costs under MI-irrigated plots.

The two determinants of the physical productivity of water are yield and irrigation water dosage, whereas the determinants of water productivity in economic terms are the gross return, input costs and the amount of water applied (Kijne et al., 2003). With the reduction in irrigation water dosage resulting from adoption of efficient irrigation technology, as seen earlier, and with a probable reduction in cost of other inputs, such as fertilizers and labour, and the enhancement of gross returns from crop produce owing to yield increase, the water productivity of the crops in both physical and economic terms changed remarkably. Comparisons show that both the physical and economic productivity are higher for MI-irrigated crops. Since we have assumed that all of the water applied to the crop is eventually depleted from the aquifer system, the improvement in applied water productivity results in real water productivity gain at the aquifer level. It can be seen that the differences in water productivity values are significant.

The cost-benefit ratio of drips and sprinklers for selected crops

In order to analyze the BC ratio of different MI systems, we have considered the major crops for which MI systems are used in the region. Although it is already known that adoption of an MI system is often associated with changes in cropping pattern (from the traditional crops to those which are amenable to MI), for our analysis we have only considered the farmers who have introduced the system without changing the crop. As a result, the values of net income used for the BC analysis will not match the net income figures shown against the same crops in Table 7.10. The reason for choosing this methodology is that it would otherwise be difficult to attribute the incremental benefits accrued after MI adoption entirely to the technology: the risks that farmers are willing to take by adopting a new crop (often a cash crop which involves market risk) would also be given credit.

Table 7.10 provides the BC analysis of nine crops irrigated by MI systems. Dhawan (2000) had earlier noted that the economic dynamic of drip irrigation is a function of the crop type, which determines the incremental income, and for high valued crops the incremental income resulting from yield improvement is likely to be very high. The BC ratio ranges from a low of 0.72 for cotton to a high of 5.93 for cluster beans. The BC ratio was second highest for fennel. These findings do not corroborate the general observations from earlier research pertaining to BC ratios for MI-irrigated crops. For instance, though cluster beans are not a high valued crop, the BC ratio is very high in this case, mainly because

Table 7.10 Benefit–cost analysis of MI systems for different crops

Season	*Crop*	*Number of observations*	*Net Income (INR/Ha)*		*Cost of WST (INR/ha/annum)*	*BC ratio*
			Before WST	*After WST*		
Kharif	Cluster bean	1	4200.00	20575.00	2763.45	5.93
	Castor	1	46500.00	57500.00	10707.79	1.03
	Groundnut	26	10415.75	28232.83	3680.93	4.89
	Cotton	1	64000.00	70200.00	8629.43	0.72
	Fennel	2	12333.33	36220.00	5512.99	5.24
Winter	Wheat	3	20922.22	53361.11	8102.00	4.49
	Potato	11	52552.08	74110.61	5556.06	4.47
Summer	Pearl millet	7	9548.57	16036.90	4396.48	2.07
	Millet	4	11856.43	22099.55	2641.02	3.71

Source: authors' own estimates based on data from primary survey.

of the low net income, without an MI system, for the only plot for which data were available, and the low capital cost of the sprinklers used for irrigating it. In that context, it is important to remember that for many crops, such as cluster beans, castor, cotton, fennel and wheat, the sample size is very small, with just one in three cases.

Having said that, the adoption of MI systems, as noted by Kumar et al., (2008) and also found in our earlier analysis for the area in question, is often associated with changes in cropping pattern. Because of this, the above analysis had limited application. It is extremely difficult to assess the economic impact of MI systems in real life situations, which are more complex. Many times, the adoption of MI goes hand in hand with a farmer's decision to introduce crops, such as groundnut, potato and chilli, which are high valued and, incidentally, very amenable to MI systems. In this case, the incremental income benefit would be much greater than our estimates. The cases in which the adopter farmers had grown the same crop prior to the adoption of MI are very rare (examples are cluster bean, cotton, millets, castor and fennel). The two exceptions are groundnut and potato.

We will see in the next section that the incremental income of the adopter farmers is very high in contrast to the not-so-impressive benefit–cost ratio of MI systems for many crops. This is because of the changes in crop composition which are not captured in the BC analysis. The adoption of certain new crops, such as fennel, pomegranate and vegetables, increases the net income substantially, but does not get captured in the BC analysis because of methodological limitations. For instance, the net return is INR 5,24,250 per hectare for chilli with micro-irrigation; INR 81,662 per hectare for pomegranate with MI; INR 52,822 per hectare for cotton with MI; but only INR 15,082 per hectare for summer bajra. Hence, the real incremental economic benefit is realized through a shift to high valued crops that give a very high return per unit of land.

The impact of adoption on overall returns from farming

For the adopters, a combination of factors can help change the overall net return from farming. They are: a) the shift in cropping pattern towards those which yield higher returns per unit area of land; b) changes in the net return from crops which are under a MI system owing to the beneficial impact of such technology (e.g. yield improvement, improvement in the quality of produce and a saving in the cost of inputs); and c) changes in livestock composition – towards those which yield higher net returns per animal – or animal holding size or an improvement the livestock rearing practices. The farmers can also increase their net returns by expanding the area under irrigation, which might be at the cost of increased groundwater use. However, this cannot be counted as the impact of the MI system or the high valued crops, as the objective of the agricultural water management interventions was to reduce the use of groundwater for irrigation. Therefore, we have considered the changes in net return per unit of land after the adoption.

Table 7.11 shows the change in the composition of the income of the adopter families before and after the adoption of MI systems. It can be seen that income from crop production had increased substantially (by INR 98,342 per annum), whereas that from dairying had gone up by INR 13,912 per annum and that from the sale of water to neighbouring farmers by INR 175. Hence, the average total increased income was INR 1, 12,429 (US$1 equals INR 50). These estimates are based on current prices and the income figures for the post-adoption scenario are not adjusted for inflation. Still, one can say that these figures are exceptionally high. Such high jumps in the annual income of a farm household can change the entire household dynamic either be positively or negatively, especially when we consider the fact that most of it is realized from select high valued cash crops like chilli, newly introduced by the farmers, which are susceptible to both production and market risks. Therefore, this aspect of the income impact needs much more careful and intensive study from a sociological angle.

Changes in overall groundwater use for farming

A major question was whether, with the reduction in the water requirement per unit of land achieved through water use efficiency improvements, the farmers would have greater incentive to expand the area under irrigation by allocating the "saved water". Peter McCornick (Director for Asia, International

Table 7.11 The impact of adoption on farm income

WST adapter	*Agriculture*	*Dairy*	*Water selling*	*Other*
Incremental benefit	207929.82	98342.11	*Water selling*	*Other*
Before	109587.72	45684.21	175.44	0.00
After	207929.82	59596.49	350.88	0.00
Incremental benefit	98342.11	13912.28	175.44	0.00

Source: authors' own estimates based on data from primary survey.

Water Management Institute, Colombo, Sri Lanka; personal communication, December 2005) argued that with higher income return from every unit of water pumped, the farmers would be tempted to invest a greater amount in tapping groundwater for growing high valued cash crops.

But field surveys showed that, in most situations, expansion of the irrigated area did not occur after the adoption of both MI systems and water-efficient crops, such as pomegranate, although the area under those crops that are amenable to MI systems or water-efficient increased. One reason for this was that the the farmers were already irrigating their entire land. In a few situations, where the land holding was large and it was practically impossible to irrigate fully due to the limited hours of power (the region is experiencing power supply rationing, with the total power supply to the farm sector limited to eight hours per day), the farmers resorted to expanding the irrigated area post-adoption of MI. However, even where this was the case, the income from farming increased remarkably. In both situations, the water productivity (INR per cubic metre) was enhanced and, in most situations, the aggregate groundwater use at the farm level reduced.

Other critiques argued that use of MI systems would only result in "applied water saving" and not "real watersaving" as, according to them, the return flows under conventional methods of irrigation would be available for reuse, and real water saving can occur only if there is reduction in crop ET. However, north Gujarat has a semi-arid to arid climate and alluvial aquifers with a deep vadose zone. In such situations, the return flows would not be available for reuse, and instead would be part of the total water depleted water consisting of "non-recoverable deep percolation" and soil evaporation (see Allen et al., 1998, for details). Hence, MI adoption actually led to real water saving at the basin/aquifer level. This was also confirmed by field investigations.

Some scholars have expressed concern that farmers in the region have limited incentives to adopt water-saving technologies under the current policy regime. The reasons they cited were that a) the water-saving and energy saving benefits from the use of MI systems do not translate into income benefits for most farmers, who are not confronted with a positive marginal cost of using water and electricity; and b) the farmers are not confronted by an opportunity cost of over-pumping groundwater (see Kumar et al., 2008; Kumar and Amarasinghe, 2009). However, the NGI interventions showed that it is possible to motivate farmers to adopt water-saving MI systems without providing subsidies, even in the absence of efficient electricity pricing in the farm sector, which can encourage efficient water use in agriculture. One strong incentive for farmers to go for MI systems was the reduction in water level "drawdown" and the consequent reduction in incidence of well failures. This was mainly because of reduction in pumping owing to improved water productivity. Another incentive was the higher yield and income they obtained post-MI adoption.

The estimates of average farm-level water use for different crops before and after adoption of a MI system show that the total farm-level water use went down from 34,870 m^3 to 27,343 m^3. The total reduction in groundwater use was 7,527 m^3. The annual saving in groundwater for irrigation was estimated to be 56.90 MCM for the current level of MI adoption.

If we assume that around 50,000 ha of the irrigated area in the alluvial parts of the region would be under MI systems, the total reduction in groundwater use would be around 112 MCM per annum. If we assume that nearly 100,000 ha of the groundwater irrigated crop inthe alluvial districts of north Gujarat (comprising Mehsana, Banaskantha, Gandhinagar and Patan), is put under MI systems, the area under MI adoption will be around 11 per cent of the well irrigated area. This is quite achievable. The water saving in that case would be around 224 MCM per annum. When compared to the total groundwater over-draft in these districts, which is 690 MCM (IRMA/UNICEF, 2001), this is a significant water saving.

Findings

Contrary to the conventional belief that water-saving MI technology adoption, which results in "applied water saving" per unit area of irrigated crop, motivates farmers to expand the area under irrigation and, accordingly achieves no real water saving at the farm level, the area under irrigation has not increased after MI adoption in north Gujarat. The area under cereals such as wheat, millet, pearl millet and rajgaro had reduced substantially with MI adoption and the introduction of high valued crops at the farm level, and is not compensated by the improvements in yield due to the use of MI systems. The reduction in cereal production can have significant implications for the domestic food security of the adopter farmers in the immediate term. More importantly, large-scale MI adoption will have serious implications for regional food security in the medium and long term.

Overall, MI technology adoption had resulted in a reduction in water application for the crops. The extent of the reduction in water application varies from crop to crop. Since all of the water applied to the crop is treated as water depletion from the aquifer, the reduction in water application can be treated as resulting in real water saving at the field level. The technology adoption had also resulted in an improvement in the yield of most of the crops covered by the technology. On an average, for most crops, the net returns from MI-irrigated plots are higher than those of plots irrigated by the conventional method, while, for the high valued crops such as chilli, the increase in income was exceptionally high.

The water productivity of the crops irrigated by MIs, in both physical and economic terms, was found to be much higher than that of their counterparts irrigated by the traditional method. The benefit–cost analysis of MI systems, for select plots-for which crop-shift has not taken place after adoption, shows significant variations in the BC ratio across crops, from as low as 0.72 to a high of 5.96. But most farmers simultaneously changed the crop with the introduction of MI system. Therefore, the analyses which consider the crop to remain the same after adoption have very limited practical and policy relevance. In real life situations, MI adoption is associated with the selection of high valued crops for which MI systems are the "best bet" technology (Kumar et al., 2008) and, as a result, the incremental benefits would far exceed our estimates. Having said that, carrying out a BC analysis of MI systems involves complex considerations of what crops farmers were growing prior to adoption, what new crops farmers chose along with the technology, and whether the risk-taking tendency of the adopter farmers is associated with confidence in precision-irrigation technology.

The overall impact of MI adoption on the income of adopter families is remarkable, crossing INR 1 lac per annum. Such high jumps in the annual income of a farm household can change the entire household dynamic. This is not necessarily always positive, especially when we consider the fact that most of it is realized from select high valued cash crops, like chilli, which are susceptible to a high degree of production and market risks.

Finally, the adoption of MI systems with the introduction of new water-efficient crops had resulted in a significant reduction in water use at the farm level. The average reduction was estimated to be 7,527 m^3 per farm whereas, at the regional level, the total groundwater saving for irrigation was estimated to be 59 MCM per annum.

Conclusions and policy

Our analyses show that adoption of MI systems is leading to large-scale effects at the farm level, from both physical and socio-economic perspectives. It is not merely that the reduction in water use is significant: the income enhancement is also quite phenomenal. Having obtained positive results from the use of MI systems for various crops, the farmers are showing an increasing preference for growing those crops, including vegetables, high valued cash crops and fruits, in place of traditional cereals. In the immediate term, this will cause a decline in cereal production affecting the domestic food security of the adopter families. But the large-scale adoption of MI systems in north Gujarat, which would eventually replace traditional cereals with high valued cash crops, can have significant implications for regional food security in the medium and long run, while creating positive impacts on the region's groundwater balance.

But a phenomenal rise in farm income can change the entire household dynamic negatively, as most of it is accrued from select high valued cash crops that pose a high degree of production and market risks. This aspect of income impact needs much more careful and intensive analysis from a sociological perspective. The domestic and local food security impacts of large-scale adoption of MI systems are a matter of concern with the increasing popularity of MI systems in several semi-arid and arid, water-scarce, regions of India, and the tendency of farmers to modify the cropping system to make it more amenable to MI.

These are major challenges for India. While improving water productivity in agriculture is extremely important for sustaining agricultural production and ensuring food security (Kumar, 2003), the technological solutions to achieve them can cause a significant negative impact on regional food security (Kumar and van Dam, 2009). But even domestic and local food security can be at risk. There are two reasons for this. First, the families will have to depend on food purchased from the market. While the quality of the commodity can be controlled by the farmers, a large reduction in cereal output can cause food shortages in the local market with a consequent increase in prices, affecting the access of local population to food and nutrition. Second, the large-scale adoption of MI systems, with associated changes in the cropping system in a region, can result in significant boost in the production of high valued crops that are friendly to MI, with a resultant drop in price of that produce (Kumar and van Dam, 2009). This, in itself, can affect the ability of the families to purchase food from the market, as their farm income can severely suffer.

Micro-irrigation for poor African countries

Unlike India, the challenge for most of sub-Saharan Africa is not of reducing consumptive water use in agriculture or to make more water available for the environment. The region has plenty of un-utilized water resources, but with very low population density, as against the physical scarcity of water in semi-arid and arid regions of India. The scarcity of water in sub-Saharan Africa is mostly due to a lack of adequate financial resources to invest in water development infrastructure and human resource capacities (Falkenmark and Rockström, 2004). Therefore, the challenge is to reduce the cost of irrigation by minimizing the amount of water required by farmers for irrigating their crops and, on the other hand, increase the returns.

How much water depleted from the system can be really saved through micro-irrigation, or what farmers do with the saved water, are of not much relevance here. Instead, the focus has to be on reducing "applied water" and energy use for irrigation, in order to reduce the cost of irrigation. At the same time, the cost of the micro-irrigation technology for saving a unit volume of water should be more than the sum of the cost saving in irrigation and the income gain due to yield benefit. Hence, two conditions will have to be satisfied. First, the system has to be technically efficient, as well as cheap. Second, the cropping system selection and agronomic inputs have to be appropriate for farmers to realize sufficient income.

During the past decade or so, a debate has been growing on promoting micro-irrigation systems as a way to improve the water security of poor farmers in sub-Saharan Africa by increasing crop production and thereby moving them out of hunger and poverty (Postel et al., 2001; Ngigi, 2008). Conventional micro-irrigation technologies with pressurized drip and sprinklers will not find many takers in most parts of sub-Saharan Africa due to poor affordability. Therefore, low cost systems such as bucket kit drips and drum kit drips are some of the technologies widely promoted by international organizations such as International Development Enterprises (IDE), Winrock International, the International Water Management Institute (IWMI), the Kenya Agricultural Research Institute (KARI) and the Horn of Africa Rainwater Partnership (GHARP) (Ngigi, 2008).

References

Allen, R. G., Willardson, L. S., & Frederiksen, H (1998). Water use definitions and their use for assessing the impacts of water conservation. In J. M de Jager, L. P. Vermes, & R. Rageb (Eds.), *Proceedings of the ICID workshop on sustainable irrigation in areas of water scarcity and drought* (pp. 72–82) Oxford, UK: International Commission on Irrigation and Drainage.

Contor, B. (2009). *Groundwater banking and the conjunctive management of groundwater and surface-water in the Upper Snake River basin of Idaho.* Idaho, USA: Idaho Water Resources Research Institute.

Custodio, E. (2000). *The complex concept of groundwater over-exploitation.* Papeles del proyecto aguas subterráneas A1. Santander, Spain: Fundacion Marcelino Botin.

Dhawan, B. D. (2000). Drip irrigation: Evaluating returns. *Economic and Political Weekly, 35*(42), 3775–3780.

Dubash, N. K (2000). Ecologically and socially embedded exchange: "Gujarat Model" of water markets. Special Article. *Economic and Political Weekly, 35*(16), 1376–1385.

Falkenmark, M., & Rockström, J. (2004). *Balancing water for humans and nature: The new approach in ecohydrology.* New York, USA: Earthscan.

Government of Gujarat. (1996). *Integrated water resources development plan for Gujarat.* Report prepared by Tahal Consultants, Gandhinagar. New Delhi, India: Government of India.

Government of India (2005). *Dynamic ground water resources of India.* New Delhi, India: Central Ground Water Board, Ministry of Water Resources, Government of India.

Institute of Rural Management Anand/UNICEF. (2001). *White paper on water in Gujarat.* Report submitted to the Department of Narmada Water Supplies, Government of Gujarat. Gandhinagar, Anand, India: Institute of Rural Management Anand.

Kijne, J., Randolph, B., & Molden, D. (2003). Improving water productivity in agriculture: Editors' overview. In J. Kijne, B. Randolph, & D. Molden (Eds.). *Water productivity in agriculture: Limits and opportunities for improvement. A comprehensive assessment of water management in agriculture.* Wallingford, UK: Centre for Agricultural Bioscience International (CABI) Publishing in association with the International Water Management Institute.

Kumar, M. D. (2000). Institutional framework for management of groundwater resources: A case study of community organizations in Gujarat, India. *Water Policy, 2*(6), 423–432.

Kumar, M. D. (2007). *Groundwater management in India: Physical, institutional and policy alternatives.* New Delhi, India: Sage Publications.

Kumar, M. D. (2009). *Water management in India: What works, what doesn't.* New Delhi, India: Gyan Books.

Kumar, M. D., & Amarasinghe, U. (Eds.). (2009). *Water productivity improvements in Indian agriculture: Potentials, constraints and prospects.* Strategic Analysis of National River Linking Project (NRLP) of India Series 4. Colombo, Sri Lanka: International Water Management Institute.

Kumar, M. D, & van Dam, J. (2009). Improving water productivity in agriculture in India: Beyond 'more crop per drop'. In M. D. Kumar, & U. Amarasinghe (Eds.), *Water productivity improvements in Indian agriculture: Potentials, constraints and prospects.* Strategic Analysis of National River Linking Project (NRLP) of India Series 4. Colombo, Sri Lanka: International Water Management Institute.

Kumar, M. D., Shah, T., Bhatt, M., & Kapadia, M. (2004). *Dripping water to a water guzzler: Techno-economic evaluation of drip irrigation in Alfalfa.* Proceedings of the 2nd Asia Pacific conference of hydrology and water resources, Suntec, Singapore, July 2004.

Kumar, M. D., & Singh, O. P. (2001). Market instruments for demand management in the face of scarcity and overuse of water in Gujarat, western India. *Water Policy, 5*(3), 387–403.

Kumar, M. D., Turral, H., Sharma, B., Amarasinghe, U., & Singh, O. P. (2008). Water saving and yield enhancing micro-irrigationtechnologies in India: When and where can they become best bet technologies? In M. D. Kumar (Ed.), *Managing water in the face of growing scarcity, inequity and declining returns: Exploring fresh approaches.* Hyderabad, India: International Water Management Institute.

Llamas, M. R. (2000). *Some ideas for discussion* (focus session on groundwater). Tenth World Water Congress. Melbourne, Autralia, March 13.

Molle, F., & Turral, H. (2004). *Demand management in a basin perspective: Is the potential for water saving over-estimated?* Paper prepared for the International Water Demand Management Conference, The Dead Sea, Jordan, June 2004.

Narayanamoorthy, A. (2004) Drip irrigation in India: Can it solve water scarcity? *Water Policy*, *6*, 117–130.

Ngigi, N. Stephen (2008). Technical evaluation and development of low-head drip irrigation systems in Kenya. *Irrigationand Drainage*, *57*, 450–462.

Palanichamy N. V., Palanisamy, K., & Shanmugam, T. R. (2002). Economic performance of drip irrigation in coconut farmers in Coimbatore. *Agricultural Economics Research Review*, conference issue, 40–48.

Postel, S., Polak, P., Keller, J. (2001). Drip irrigation for small farmers: A new initiative to alleviate hunger and poverty. *Water International*, *26*(1), 3–13.

Ranade, R., & Kumar, M. D. (2004). Narmada water for groundwater recharge in north Gujarat: Conjunctive management in large irrigation projects. *Economic and Political Weekly*, *39*(31), 3498–3503.

Rosegrant, M. W., & Schleyer, R. G. (1996). Establishing tradable water rights: Implementation of the Mexican water law. *Irrigation and Drainage Systems*, *10*(3), 263–279.

Saleth, R. Maria (1997). Power tariff policy for groundwater regulation: Efficiency, equity and sustainability. *Artha Vijnana*, *39*(3), 312–322.

Shah, T., Deb Roy, A., Qureshi, A. S., & Wang, J. (2003). Sustaining Asia's groundwater boom: An overview of issues and evidence. *Natural Resources Forum*, *27*, 130–141.

Singh, O. P., Sharma, A., Singh, R., & Shah, T. (2004). Virtual water trade in the dairy economy: Analyses of irrigation water productivity in dairy production in Gujarat, India. *Economic and Political Weekly*, *39*(31), 3492–3497.

Suresh Kumar, D., & Palanisami, K. (2011). Can drip irrigation technology be socially beneficial? Evidence from southern India. *Water Policy*, *13*(4), 571–587.

Thobani, M. (1997). Formal water markets: Why, when and how to introduce tradable Water rights. *The World Bank Research Observer*, *12*(2), 161–179.

Zekri, S. (2008). Using economic incentives and regulations to reduce seawater intrusion in the Batinah coastal area of Oman. *Agricultural Water Management*, *95*(3).

8 Technology choices and institutions for improving the economic and livelihood benefits from multiple-use tanks in western Odisha

M. Dinesh Kumar, Ranjan Panda, V. Niranjan and Nitin Bassi

Introduction

Eastern India has the largest concentration of population in the world and also houses the largest number of the world's poor. It suffers not only from economic scarcity of water, but also contains many small holdings and has a high degree of land fragmentation (Kumar et al., 2009). Owing to poor rural electrification, well owners in the region use diesel engines for water abstraction and spend large sums for irrigation, whereas the non-well owning farmers buy irrigation water at exorbitant prices (Kumar, 2007; Shah, 2001). Landlessness is also a major problem. The region's rural economy is purely agrarian with paddy as the main crop. But its agriculture suffers from low productivity, owing to the low level of adoption of agricultural technologies, the high cost of irrigation water, social and ecological problems and poor rural infrastructure.

The region's landscape is dotted with numerous small tanks, which are under the common property regime and governed by the Panchayats. Government agencies recognize and also operate minor irrigation systems as single-use systems. However, these tanks are also used for fishing, as a source of both water for domestic needs and nutrient rich soils, and for fodder grass collection and brick making. These uses have high value in terms of household income, nutrition and health, for the poorest of the poor. Owing to this lack of recognition, water from MI tanks gets diverted for irrigating low valued crops.

Well-designed multiple-use systems (MUS), which involve integrating fisheries, prawn farming and duckkeeping with paddy irrigation, using local secondary reservoirs for the water, can enhance the productivity of the use of both land and water in eastern India to a considerable extent (Sikka, 2009). Research in south India shows how revenue maximization can be made possible by using irrigation tanks for multiple uses, such as social forestry, brick making, fisheries, silt collection and groundwater recharge (Palanisami et al., 2010).

Tanks and ponds have been the primary source of water for poor rural households for domestic use, irrigation and fish production in Odisha. Generally, neither the Minor Irrigation Department nor the Panchayats have designed the infrastructure for multiple uses, but the system, by default, becomes a multiple-use system. As noted by van Koppen et al. (2009), while some of the unplanned uses may get absorbed by the system, other uses can damage it. But, these tanks/ponds are an important source of drinking water for poor rural households. Thus, these water systems are characterized by competing water needs, and are under severe stress. In the process, they are kept out of reach of the poor.

The main reason is that the governance of these common property tanks is either poor or is totally absent. Even when a governance system for tanks and ponds exists at the local level, there is a lack of clarity on the legitimate uses of these tanks; the rights owners; who should manage them; and what should be the role of the local community in the management. Wherever local management institutions exist, are either not capable of allocating water from these systems to meet multiple demands, or are not mandated to do so. It is also important to recognize that communities often do not realize the costs of using these tanks for certain purposes.

In the absence of good governance, the right to water from these tanks is often politically contested. The more politically powerful and socially-dominating groups often take control of these systems for commercial activities, at the cost of the rights of drinking water users or, sometimes, the local fishing community. This leads to their sub-optimal performance from a social, environmental and even economic angle. Such actions leave minimum incentives for community members to manage them, leading to further degradation. But creating institutions for governance alone would not be sufficient to make these water systems into efficient and effective multiple-use systems. This is because the physical condition of these systems is also influenced by "negative externalities", such as annual variations in climate, and even catchment/basin management decisions, which are beyond the "sphere of influence" of local tank institutions. These factors can weaken the performance of the local institutions.

Hence, the following steps are required. The first step is improvement in the physical system, in order to ensure enhanced quality and reliability of the water so as to meet basic needs. The second step is the institutional innovation that ensures the hydrological integrity of the local MUS, management of inter-sectoral water demands, and access to water for the poor. One important feature of this innovation would be coordination of local tank management decisions (and basin-level management decisions for large systems). But lack of proper quantitative understanding of the costs and benefits associated with multiple uses inhibits public investment for these innovations.

The questions that need to be addressed are as follows. How reliable are the multiple-use systems for provision of water for basic needs and productive uses? How equitable is the distribution of water from these systems across use sectors? How does access equity change with variations in climate? How optimal is the allocation of water from these systems from economic, social and environmental points of view? What physical system improvements are possible for enhancing the overall performance of these systems from economic, environmental and social points of view? What kind institutional innovations are needed to effect these changes?

General features of western Odisha

The net irrigated area from various sources in Odisha stood at around 1.3 million hectares in 2003–04. Of this, irrigation by canals was 0.9 million hectares and irrigation by tanks was around 0.102 million hectares (source: Indiastat). Nearly 40 per cent of the total minor irrigation (MI) schemes are also located in southern and western parts of Odisha, which have traditional tanks (ADB, 2006). The state has about 28,303 tanks with a potential to irrigate about 0.69 million hectares. Of these, 3,847 tanks are relatively large with an irrigation capacity of 5.69 lac ha. The large tanks have a command area of between 40 ha and 2,000 ha, and are managed by the Minor Irrigation Department.

The net irrigated area by tanks in Odisha during 1950–51 was about 5.46 lac ha, which was about 54.22 per cent of the total net irrigated area in the state. From 1956 onwards, the tank-irrigated area started declining over time in absolute figures and not just in terms of proportion to the total net irrigated area (ADB, 2006). However by 2003–04, the net area irrigated by tanks in the state came down to 1,02,000 ha, which was only 7.7 per cent of the total net area irrigated. This decline was mainly due to the poor maintenance and management of the tank systems.

The Sambalpur district of western Odisha lies between 20° 40'N and 22° 11'N latitude, 82° 39'E and 85° 15'E longitude, with a total area of 6,702 km². In this region, the rural landscape is dominated by tanks of various sizes. They form a major source of irrigation for the poor small and marginal farmers of the region. The only other source of irrigation is the limited groundwater, which is available in the crystalline formations at a depth of nearly 30–40 feet, can be tapped efficiently only through open wells (GOO, 2007). But the cost of digging a well is very high,[1] and the poor farmers in the area cannot afford this. Consequently, very few farmers in the area have open wells. All of these factors make tanks very important for sustaining irrigated agriculture, and thus the livelihoods of poor farmers, in the area.

The district has three distinctive physiographic units, examples of which are the hilly terrain of Bamra and Kuchinda in the north, the plateau and ridges of Rairakhol in the south-east and the valley and plains of the Sambalpur sub-division in the south-east. Sambalpur district experiences an extreme type of climate with 66 rainy days and 1,530 mm rainfall on average in a year. Most of the rainfall is confined to the months from June to October when the district is visited by the south-west monsoon. The mercury rises up to 47°C during May, when there is a heat wave, and falls as low as 11.8°C during December, when there is extreme cold. The rainfall is highly uneven and erratic.

The dominant soil found in the district is light red laterite, which has a high clay content. The soil depth ranges from 0–22 cm. The soil belongs to the texture class of sandy clay loam (Sahu and Nanda, undated). They have low infiltration rates when thoroughly wetted – namely 0.17 inches/3.8 mm per hour (Texas Council of Government, 2003, as cited in Oram, 2009) – and consist chiefly of soils with a layer that impedes the downward movement of water and those soils with a moderately fine to fine structure.

The district forms a part of the Mahanadi River basin. The Mahanadi, the largest river of the state, with a total drainage area of 1,43,000 km², enters into the district in the north-western border, where the famous Hirakud multi-purpose reservoir project is situated. The flows into the Mahanadi basin constitute the largest amount of surface water among all river basins in the state of Odisha. The annual flow (at 75 per cent dependability) of the river at the Hirakud dam site, with a total upper catchment area of 83,400 km² is 24.853 BCM (GOO, 2007). The mean annual rainfall, potential evaporation and run-off in the basin upstream of the Hirakud dam are given in Figure 8.1 (source: authors' own estimates based on GOO, 2007). From these figures, it can be inferred that August is the wettest month of the year in terms of surface water flows, as a result of high rainfall and low evaporation.

The objectives

The specific objectives of the research study were to:

- Analyze the various existing demands for water – for both consumptive and productive water needs – from the individual households and the community at large.
- Analyze how various tank uses, and the degree of equity in access to water, change with drought and floods, and their likely impact on the livelihoods of the poor.
- Assess the economic value of the various benefits and costs associated with different tank uses, and how they change in response to climatic variability.
- Analyze the trade-off between maximizing the direct economic outputs, and optimizing the economic, social and environmental benefits as well as the poverty reduction impacts of MUS systems.

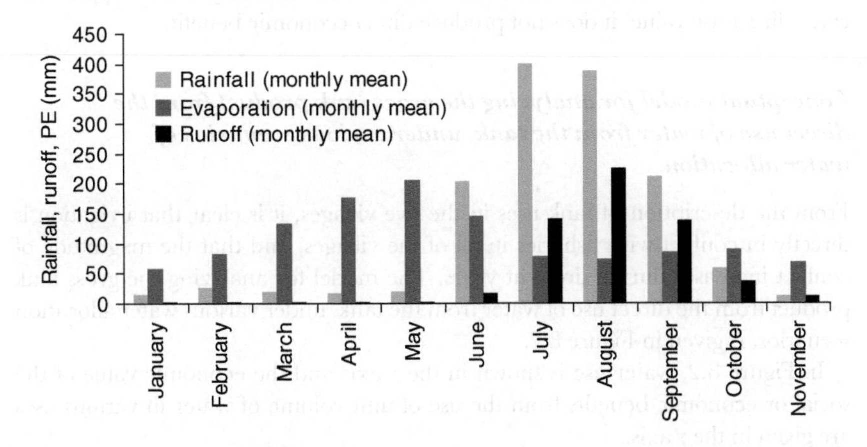

Figure 8.1 Monthly rainfall, potential evaporation and runoff (mm) in Mahanadi basin, upstream of Hirakud.

Source: authors' own estimates based on Government of Odisha, 2007.

- Assess the physical improvements in the tank system for improving their overall performance as MUSs.

Approach, methods and tools

The approach

First, the total value of the existing uses of the wetland will be assessed, including the value of the (direct and indirect) economic, social and environmental benefits (Turner et al., 1998). Then various scenarios for enhancing the direct economic output from the tank will be generated (with different degrees of constraint induced from the point of view of social and environmental sustainability) and the value of the economic benefits will be compared, along with and the total economic value of all associated benefits (social, economic and environmental) produced by the wetland ecosystem under each scenario. This forms an integrated ecological-economic modeling. This model can be characterized by simple theoretical models that aggregate (Costanza et al., 1993). This can provide indications as to what extent the social and environmental values could be compromised in order to realize higher economic returns from the use of tanks.

Here, our basic premise is that there is always a conflict between maximizing the direct economic benefits from the tank uses and ensuring the sustainability of the social and environmental benefits of the tank ecosystem. For instance, intensive fishery in the tank would reduce make the tank water unusable for drinking purpose due to the use of fish feed and fertilizers. Similarly, allowing animal grazing in the tank bed during dry season would lead to the deposition of animal waste and compaction of the tank bed, thereby reducing rainwater infiltration into the soils and aquifer recharge. Here, it is important that we do not confuse the direct economic benefits from the direct use values. While drinking water supply concerns direct use value, it does not produce direct economic benefit.

Conceptual model for analyzing the gross tank product from the direct use of water from the tank, under various scenarios of water allocation.

From the description of tank uses in the five villages, it is clear that irrigation is directly in conflict with fisheries in all of the villages, and that the magnitude of conflict increases during drought years. The model for analyzing the gross tank product from the direct use of water from the tank, under various water allocation scenarios, is given in Figure 8.2.

In Figure 8.2, water use is shown in the *x* axis and the economic value of the social or economic benefits from the use of unit volume of water in various uses are given in the *y* axis.

In all situations, water has to be kept for domestic use and livestock drinking. This is non-negotiable, as the social values generated from these are significant. We have converted this into monetary values by considering the cost incurred by the public

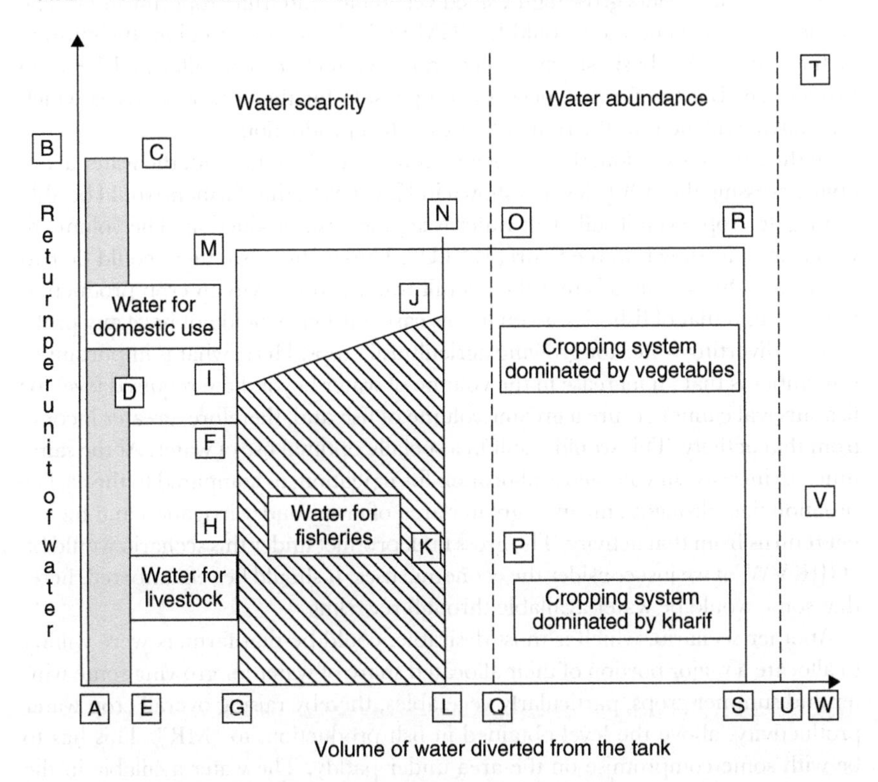

Figure 8.2 Conceptual model for analyzing surplus value product from the direct use of tank water under various scenarios of water allocation.

Source: authors' own conceptual model diagram.

utility in creating a similar facility. Thus, the values generated from these two services are expressed in economic return per unit of water. In the case of fisheries and crop production, it is expressed in net income return per unit volume of water used, or net water productivity (INR per cubic metre (US$ 1 is INR 50)).

Two typical situations are possible with regard to water availability. First, the water available in the tank is just sufficient for fish production. The total water that can be diverted from the tank would stand at the level shown as "OQ" in Figure 8.2. Let us assume a scenario in which the water productivity in fisheries is higher than the overall irrigation water productivity for crop production. In that case, the economic return from fisheries and some irrigated crops (the area embedded in "GIJKPQ" in Figure 8.2) will be higher than that which can be derived from irrigated crop cultivation alone ("GHPQ"). Under such a scenario, fisheries need to be encouraged, provided that a share of the benefits from them go to large number of the farmers. In the next scenario, let us assume that the overall irrigation water productivity in crop production is higher than that under fish production. This is possible when farmers grow high valued vegetables and fruit crops in winter. The economic value generated would be "GMNL".[2] In such a situation, the entirety of the water (after basic survival needs and livestock) can be allocated for crop production. But the farmers need to compensate for the economic losses which Panchayat will incur as the result of a loss of fish production.

In the second situation, there is plenty of water in the tank (with the water availability crossing the "OQ" level as shown in Figure 8.1); thus farmers would be able to irrigate crops even if sufficient water is kept for fish production. The volume of water here, as shown in the figure, is "TU". Under this, also, there could be two scenarios. One scenario is that the overall water productivity in crop production is lower than that of fisheries. Even in this case, farmers would not have any problems in diverting water for growing agricultural crops. Here, what is important to remember is that an increase in the volume of water beyond the required level for fish survival cannot ensure a greater volume of fish and, therefore, greater income from that activity. This would result in a socially optimal use of water. At the same time, an increase in volumetric allocation for irrigation, as compared to the earlier situation (i.e., drought), means a greater area of crops under irrigation and higher net returns from that activity. The gross tank product under this scenario would be "GIJKVW" if we just consider the economic uses. It should be remembered, here, that some would be water available through recycling.

Another scenario, which is most desirable, could be that farmers were willing to allocate a major portion of their allocated share of water for growing some winter and summer crops, particularly vegetables, thereby raising overall crop water productivity, above the level obtained in fish production, to "MR". This has to be with some compromise on the area under paddy. The water available in the tank for crop production would be only at the "RS level", as some water would be evaporated while being kept in storage for winter and summer use. Here the gross tank product would be "GIJNRS". In this case, again, compromising on fish production would mean greater economic return from tank water use, i.e., the area shown by "GMRS" in the diagram. But this requires that the fishing community or the Panchayat, which leases out the tank to a fishing contractor, is compensated for the revenue losses, i.e., the area under "GIJL". But this would still give greater income to the farmers, to the tune of "IMNJ".

However, in the entire analytical framework, we have not considered the potential variations in water productivity and surplus value product generated from the unit volume of water for crop production between the years of water scarcity and water abundance. This is mainly because years of water scarcity coincide with years of meteorological droughts, in which crops demand a higher quantum of irrigation water.

Using primary data collected from the five tanks studied, we first analyze the gross value generated from the current tanks' uses under the two different situations of water availability (a drought year and a good rainfall year) based on the volume of water diverted for various uses and the overall net water productivity secured under those uses. If less water is allocated to uses that have the capacity to absorb more water and give high returns per unit volume of water, then existing water allocation can be said to be sub-optimal. Subsequently, we examine whether there is scope for reallocating water across use sectors and sub-sectors, in order to enhance the surplus value product generated based on economic efficiency considerations. The basic premise is that more water will be allocated to uses that generate higher economic value. The data for the same will be obtained from water productivity estimates for various crops and fisheries. However, the net returns from fisheries are assumed to be constant, as an increase in the water available in the pond for fish production would not result in increased returns from the same.

Analytical procedure for wetland valuation

The total economic value (TEV) of tanks in Odisha can be assessed by taking the sum of the net economic return from the use of tank water for irrigation (DUB); the net economic return from fish catch (DUB); the economic value of the drinking and domestic water use benefits (DUF); the economic value of the benefits produced by water supply to livestock (DUB); and the recreation value of the tank (IUB). The benefit of nutrient-laden soils was not considered in estimating total economic value, as people were not found to be using the silt from the tank bed. Also, the benefit of tank bed cultivation was not considered, as tank bed cultivation is not a practice in the area.

The economic value of irrigation water ($EV_{IRRIG\text{–}WATER}$) can be estimated by taking the ratio of the incremental net income per unit area of that land which is irrigated by tank water alone, over that which is irrigated using other water sources or unit rain-fed area (the hedonic pricing method) and the average volume of water used per unit area of irrigation, and multiplying it with volume of water used for irrigation from the tank ($V_{IRRIGATION}$). Here, it is assumed that the farmers who use other sources of water would eventually incur higher costs for irrigation as compared to the tank users, and, on the other hand, that the tank users would get higher returns for the same level of inputs, by virtue of the presence of micronutrients in the water.

$$EV_{IRRIG-WATER} = \frac{V_{IRRIGATION} X (ANR_{IRRIGA-CROP} - ANR_{RAIN-CROP})}{AV_{UNIT-LAND}} \quad (1)$$

Here, $AV_{UNIT-LAND}$ is the average volume of water required to irrigate a unit area of crop land irrigated with the use of tank water. This can be estimated for the existing irrigated cropping pattern in the area using the estimated values of the depth of irrigation for different crops, and the area under each crop. This also means that, just by manipulating the cropping pattern, the $EV_{IRRIG-WATER}$ can be enhanced. It is important to note that the values in brackets, when divided by $AV_{UNIT-LAND}$ give the overall net water productivity of the crops irrigated by the wetland.

$$ANR_{\text{IRRIG-CROP}} = \sum_{i=0}^{n} (\text{NR}_i) A_i \Big/ \sum_{i=0}^{n} A_i \tag{2}$$

Here, NR_i is the net return from crop *i*; and A_i is the area under crop *i* from the sample farmers, and *n* is the total number of crops grown by the farmers in the tank command.

The recreational value of the tank ($EV_{\text{RECREATION}}$) can be assessed by the price people are willing to pay for availing themselves of a similar service elsewhere (such as the use of swimming pools and fishing lakes) and the total number of people using the tank for recreational purposes (hedonic pricing method) at present. Here we are not considering the bio-diversity value of the tank, as the tanks are quite small, with their water spread area in the range of 1–5 acres.

The economic value of the social benefits produced by drinking and domestic water supply and the water supply available for livestock from the tank (sum of $V_{\text{DOMESTIC}} + EV_{\text{LIVESTOCK-WATER}}$) can be estimated by taking the public investment required for creating a source of water supply for the same population as that to which the tank caters. After Turner et al. (1998), this is known as the public pricing method.

$$\textbf{TEV} = EV_{\text{IRRIG-WATER}} + EV_{\text{FISH-CATCH}} + EV_{\text{DOMESTIC}} + EV_{\text{LIVESTOCK-WATER}} + EV_{\text{RECREATION}} \tag{3}$$

Constraints induced by sustainability

A minimum volume of water from the tank will have to be earmarked for domestic purposes, including human consumption and animal drinking.

The area surrounding the tank should not be used for irrigated paddy production with fertilizer use during kharif season, as the field runoff containing fertilizer and pesticide residues would contaminate the tank water, and deep percolation of water from the field would contaminate the groundwater. However, it can be allowed during the winter season when the water table drops, or during kharif if fertilizers are not used.

Intensive fish farming would contaminate groundwater with fertilizers and, therefore, would make groundwater unsuitable for drinking.

The silt from the tank should be scraped every year. Tank bed grazing should not be allowed as it would compact the tank bed, stopping natural infiltration and increasing the chances of the growth of weeds in the tank bed.

Volume of water available for irrigation and fish production

$$= V_{\text{FISH}} + V_{\text{IRRIGATION}} = V_{\text{TANK}} - V_{\text{DOMESTIC}} \tag{4}$$

$$V_{\text{DOMESTIC}} \geqslant POP * DWR_{\text{PER-CAPITA}} \tag{5}$$

$$V_{\text{TOTAL}} \leqslant V_{\text{TANK}} \tag{6}$$

But, V_{TOTAL} can be defined as:

$$V_{\mathrm{TOTAL}} = V_{\mathrm{FISH}} + V_{\mathrm{IRRIGATION}} + V_{\mathrm{DOMESTIC}} + V_{\mathrm{LIVESTOCK-WATER}} + V_{\mathrm{IN-STREAM}} + \ldots \quad (5)$$

Hence:

$$V_{\mathrm{FISH}} + V_{\mathrm{IRRIGATION}} + V_{\mathrm{DOMESTIC}} + V_{\mathrm{LIVESTOCK-WATER}} \leqslant (V_{\mathrm{TANK}} - V_{\mathrm{IN-STREAM}}) \ldots\ldots \quad (6)$$

Here, V_{DOMESTIC} is the total consumptive water use for domestic purposes; and $V_{\mathrm{IN-STREAM}}$ is the total amount of water required to be kept in the tank for in-stream uses such as washing, bathing and swimming. This would be same as the water required for recreation.

The sample design and size

A total of five tanks located in the Sambalpur district of western Odisha were chosen for detailed study: Gadloisingh, Jhankarpalli, Laida, Rengloi and Rugudipali. The study included use of primary survey of various tank users (wetland) and the farmers in the upland. A total of 240 HHs from the wetland and 240 HHs from upland were surveyed from the five tank commands. In addition, village level data were collected on the physical characteristics of the wetland (area, depth, area irrigated in normal and drought years, number of families depending on the tank for various uses, etc.). In addition, secondary data on the physical features of the region were also obtained from published and grey literature.

Multiple water use benefits

The multiple water use benefits identified from the five tanks studied were irrigation, fish farming, water for domestic use (washing, bathing, cleaning of utensils) and water for livestock drinking. In addition, recreational use of water took place in all of the tanks. It was found that the irrigated area shrinks during drought years in all the tanks. Contrary to what was found in the case of irrigation, fish production is not compromised in the years of drought. In this section, we discuss the extent of these benefits, vis-à-vis the area irrigated during normal and drought years, the crops irrigated, the yield of the crops irrigated by tank water and the incremental net return from these crops over the upland crops irrigated from other sources; the total value of the economic benefits generated from the use of tank water for irrigation; the number of families depending on tank water for domestic uses; the number of families using the tank water for livestock uses; the value of the social good produced from the use of water for domestic and livestock drinking purposes; and the quantum of fish production from the tanks and the income earned from the sale of fish.

In addition, the estimates of the total volume of water diverted from the tank for irrigation during normal and drought years, and the water productivity of crops irrigated by tank water during those years are also presented and discussed in this section.

The irrigation water use benefits

The benefits of irrigation using tank water depend not only on the total area irrigated, but also on the net return from the crops grown over the net returns if the same crops are grown without tank water or with water from an alternative source. If land is available in plenty in the area and water is scarce, farmers would be able to maximize the economic returns from irrigation by allocating water to crops that give higher returns per unit volume of water. If, on the other hand, water is available in plenty in the tank, and land is scarce in the area, then farmers would be able to maximize their returns by allocating their water to crops that give high returns per unit of land (i.e., high land productivity).

The following table gives the cropping pattern of the tank water users. This, however, only provides details of crops which are irrigated using tank water. In Table 8.1, the figures for the total area irrigated by the sample farmers (40) in the tank command are given. It is quite possible that, in certain cases, the farmers belonged to more than one tank command. As a result of this, the total area irrigated under different crops might turn out to be more than the actual area under command of the tank chosen for the study. It can be seen from the table that, in the case of Tank 2 (Gadloisingh) and Tank 4 (Rengloi), there are many winter crops (vegetables) grown with the tank water, whereas in the case of Tank 1, Tank 3 and Tank 5, only kharif paddy is irrigated with tank water. For drought years, also, a similar pattern was seen in all the five tanks studied (Table 8.2).

The yield and net return from various crops irrigated by tank water are given in Table 8.3. For all the crops, the yield figures under tank irrigation are found to be much higher than those when they are cultivated upland. The only exception is in the case of wetland paddy in Jhankarpalli. Here, the farmers do not seem to be irrigating the crop adequately, and as a result of which the yields are found to be

Table 8.1 The area under different crops irrigated from tank water in five selected villages during a normal year

Sr. no.	*Name of tank*	*Area irrigated*	*Irrigated area under crop (acre)*						
			Paddy	*Brinjal*	*Mustard*	*Onion*	*Potato*	*Tomato*	*Others*
1	Gadloisingh	Wetland	63.6						
		Upland	236.3						29.7
2	Jhankarpalli	Wetland	117.2	138.7	1.0	74.6	23.3	117.2	67.1
		Upland	164.6	1.6		17.6	17.6	0.1	16.3
3	Laida	Wetland	50.6						
		Upland	112.7						
4	Rengloi	Wetland	47.0		43.0		1.0		26.4
		Upland	113.3						
5	Rugudipali	Wetland	28.0						
		Upland	60.8						1.0

Source: data from primary survey.

Table 8.2 The area under different crops irrigated from tank water in five selected villages during a drought year

Sr. no.	*Name of village*	*Area irrigated*	*Irrigated area under crop (acre)*						
			Paddy	*Brinjal*	*Mustard*	*Onion*	*Potato*	*Tomato*	*Others*
1	Gadloisingh	Wetland	31.8						
		Upland	300.3						28.7
2	Jhankarpalli	Wetland	1.3	120.1	1	81.2	24.4	157.0	71.2
		Upland	3						
3	Laida	Wetland	54.7						
		Upland	112.7						
4	Rengloi	Wetland	49.8		43.0		12.2		16.7
		Upland	109.3						
5	Rugudipali	Wetland	24.1						
		Upland	66.6						1.5

Source: data from primary survey.

Table 8.3 Yield of crops irrigated by tank water in the five selected villages during a normal year

Sr. no.	*Name of tank*		*Yield of irrigated crops under tank command (kg/acre)*						
			Paddy	*Brinjal*	*Mustard*	*Onion*	*Potato*	*Tomato*	*Others*
1	Gadloisingh	Wetland	1158						
		Upland	1027						67
2	Jhankarpalli	Wetland	754	478	400	427	250	651	
		Upland	3141	120		201	234	250	340
3	Laida	Wetland	1321						
		Upland	1015						
4	Rengloi	Wetland	1191		417		225		
		Upland	1080						
5	Rugudipali	Wetland	1266						
		Upland	861						50

Source: authors' own analysis of primary data.

low. Farmers in this tank command seem to be keen to grow vegetables, which is evident from the fact that five different vegetables are grown by the farmers in the tank command. An important observation is that the upland farmers in Rengloi are not growing vegetables in the normal years. A similar trend was also seen during the drought year (Table 8.4). It is important to note that, during drought years, the upland farmers in neither of the two villages of Jhankarpalli and Rengloi grow the vegetables which the farmers in the tank command grow. This highlights the importance of tanks during the drought years.

The net return from irrigated crops in the tank command and those crops raised in the upland are given in Table 8.5 and Table 8.6 for the normal year studied and

Table 8.4 Yield of crops irrigated by tank water in the five selected villages during a drought year

Sr. no.	*Name of tank*		*Yield of irrigated crops under tank command (kg/acre)*						
			Paddy	*Brinjal*	*Mustard*	*Onion*	*Potato*	*Tomato*	*Others*
1	Gadloisingh	Wetland	1008						
		Upland	516						
2	Jhankarpalli	Wetland	238	422	400	432	275	486	
		Upland	300						
3	Laida	Wetland	865						
		Upland	533						
4	Rengloi	Wetland	956		400		200		
		Upland	628						
5	Rugudipali	Wetland	980						
		Upland	459						15

Source: authors' own analysis of primary data.

Table 8.5 Net returns from crops irrigated by tank water in the five selected villages during a normal year

Sr. no.	*Name of tank*		*Net return from irrigated crops under tank command per acre (INR/acre)*						
			Paddy	*Brinjal*	*Mustard*	*Onion*	*Potato*	*Tomato*	*Others*
1	Gadloisingh	Wetland	7194						
		Upland	6501						600
2	Jhankarpalli	Wetland	5367	4920	8600	5585	1963	8289	1925
		Upland	2750	1140		5800	3450	3050	3450
3	Laida	Wetland	6933						
		Upland	3769						
4	Rengloi	Wetland	7402		19215				6400
		Upland	6440						
5	Rugudipali	Wetland	7732						
		Upland	4215						

Source: authors' own analysis of primary data.

drought year respectively. While it is the case that, for paddy, the net return from the tank irrigated field was found to be higher than for that in upland (except for Jhankarpalli, which is perhaps due to the low yields; see above), for potato, onion and "other crops" the net returns were lower for farmers in the tank command. In the case of Rengloi, the only crop that the upland farmers grew both in normal and drought years was paddy. In the case of Jhankarpalli, the upland farmers were found to be growing certain vegetables just in normal years.

Based on the figures of net return from different irrigated crops, and the net return from the upland crops, and the cropping pattern found for sample

Table 8.6 Net returns from crops irrigated by tank water and, also, upland in the five selected villages during a drought year

Sr. no.	*Name of tank*		*Net return from irrigated crops under tank command (INR/acre)*						
			Paddy	*Brinjal*	*Mustard*	*Onion*	*Potato*	*Tomato*	*Others*
1	Gadloisingh	Wetland	5894						
		Upland	1972						
2	Jhankarpalli	Wetland	790	4195	12200	5565	1619	5945	1707
		Upland							
3	Laida	Wetland	2624						
		Upland	1246						
4	Rengloi	Wetland	5358		17740				4900
		Upland	2237						
5	Rugudipali	Wetland	4943						
		Upland	1621						21700

Source: authors' own analysis of primary data.

farmers in wetland and upland, we have estimated the average net return per unit area of the wetland and the upland. From these, the incremental net return for wetland irrigated land was estimated for all the five tanks. This is denoted by $ANR_{IRRIGA-CROP} - ANR_{RAIN-CROP}$ (see Equation 3).

Based on the average depth of irrigation worked out for different crops in the tank command and the cropping pattern arrived at for the sample farmers in the tank command, we have also estimated the average depth of irrigation per unit of tank irrigated area for each of the five tanks. This is denoted by $AV_{UNIT-LAND}$ in the same equation. The ratio of the first variable (in the numerator) and the second variable (in the denominator) gives the surplus value product from irrigation per unit volume of tank water. Subsequently, using the total area irrigated by the tank in normal and drought years and the average depth of watering for each crop, the total volume of water diverted from the tank can be calculated. The multiple of this with the earlier variable (surplus value product from unit volume of water) yields the total economic value generated from the use of tank water for agriculture. It is important to mention here that the estimation of the gross irrigated area by tank is somewhat tentative. We have actually treated the primary data from villagers on tank irrigated area as the area irrigated under paddy, and then worked out the area under other irrigated crops in the tank command purely based on the proportion of the area under those crops obtained from the sample farmers' data. The area reported for drought years is also tentative. For instance, in the case of Rugudipali, the reported area for the drought year was only 100 acres, whereas the same for the normal year was six times higher.

The estimates of the economic value of water in irrigated production in the tank command are given in Table 8.9. What is interesting to note is that the incremental return from irrigation per unit irrigated area is high during drought years in three out of the five tanks. This basically shows that value of irrigation water becomes critically important during drought years. During the drought years, the upland farmers are not able to secure good yields for paddy, whereas the farmers in the wetland are able to secure good yields with the availability of irrigation water.

Another interesting phenomenon is that the amount of water used during drought years is much higher than that used in normal years in three out of the five tanks. The reason for this is that a lot of the crop water demand is from paddy, and, for this crop, the water demand is being met from the rainfall. Columns 6 and 7 in Table 8.7 show this. The average depth of watering per unit area of irrigated crop is less during the normal year. During normal years, the irrigation water demand for winter crops would also be generally low. This reduces the overall water demand and water withdrawal during normal years. It is important to remember that there isn't much scope for expanding the command area of a tank during a particular season, as this is determined by considerations such as topography. Further, the restriction on water withdrawal during the winter season also limits the volumetric water use during normal years.

Nevertheless, the economic value of crop outputs produced with the use of irrigation water is much higher during normal years for four out of the five tanks. Only in the case of the Gadloising tank do the crop outputs produced during a drought year have a much higher value than those produced during a normal year. This does not mean that drought is more desirable than a normal year, in terms of income generation. It only means that the tank water has a higher value, in economic terms, during a drought year as, during the normal year, the farmers in the upland also derive sufficient income from crops which are rain-fed. In absolute terms, the poor farmers who are dependent on tanks for irrigating their kharif paddy suffer during drought years. In fact, the economic returns from tank irrigated crops would have been much higher during the normal year if water allocation had been judicious. If the farmers are able to expand the area under irrigation during a normal rainfall year, then, by building water conveyance infrastructure, the economic value of the returns would be higher during the normal year as well.

It can also be seen that, in the case of Laida, the farmers do not use irrigation water for paddy, which is the only crop grown in the tank command during normal rainfall years. This is quite possible for wetlands. The reasons are twofold: a) the wetlands in the downstream of the tank receive excessive seepage from the tanks,

Table 8.7 The total economic value of wetland irrigation from the five selected tanks

Tank Name	$ANR_{IRRI-Water} - ANR_{Rain-Crop}$		$V_{IRRIGATION}$		$AV_{UNIT-LAND}$		$EV_{IRRIG-WATER}$	
	Normal year	*Drought year*	*Normal year*	*Drought year*	*Normal year*	*Drought year*	*Normal year*	*Drought year*
Gadloisingh	1351.87	4094.03	5518.00	17108.00	27.59	171.08	270374.0	270374.0
Jhankarpalli	2254.14	4222.87	8763.90	7095.20	63.40	102.84	311594.0	291346.8
Laida	4309.00	2523.00	0.0	16770.00	0.00	167.74	269958.0	252239.8
Rengloi	5000.37	7803.46	13413.0	4879.70	107.30	195.20	625069.6	195074.5
Rugudipali	3585.20	2879.73	5211.60	14968.00	8.68	149.68	2152607.0	287973.0

Source: authors' own analysis based on outputs presented in Tables 8.3 and 8.4 and other analyses

and therefore remain wet during monsoon season and even during winter; and b) the region receives very high rainfall, which is adequate for kharif paddy (provided that the monsoon does not fail). But the problem is that it becomes difficult for the farmers to take water to areas which are actually outside the tank command, due to lack of infrastructure for water lifting and conveyance.

Fisheries production in the tanks, and its value

The economic value of annual returns from fish farming in the five tanks is given in Table 8.8. This is estimated by multiplying the total quantum of fish caught annually (kg) and the market value of the particular variety of fish per kg (the price which the consumers have to pay to get the fish from the market). In fact, all these tanks are leased out by the respective village Panchayats to fishing contractors on annual leasing. All the investment is made by the fish contractors, and they do the harvest and retain the profits. Only the lease charges are paid to the Panchayat.

The total economic value of the fish catch made by community members ranged from INR 18,400 for Gadloisingh and INR 18,300 for Rengloi, down to INR 4,400 in the case of Rugudipali.

Domestic water supply from the tank and the value of the social good

The estimates of the number of families using the tank for domestic and livestock drinking are presented in Table 8.9. In all except the fifth tank, a significant number of families depend on the tank water for domestic and livestock drinking uses other than human drinking. It is important to note that, during drought years, dependence on the tanks was reported to be low at least for two tanks. The number of families depending on the tanks for domestic and livestock uses is much higher than those who depend on them for irrigation. For instance, in the case of Jhankarpalli, 50–60 HHs depend on it for irrigation, whereas 180 families depend on it for domestic and livestock uses. This is a very significant number, highlighting the importance of the tank in the village socio-economic dynamic. Similarly, in the case of Rengloi, a total of 75 families depend on the tank for irrigation, whereas the number of families depending on it for domestic uses is 134. In the case of Laida, a total of 48 families were reported to be using the tank for irrigation, whereas a total of 500 families depend on it for domestic uses in a normal year. The number comes down to 413 during a drought year.

In order to estimate the value of these services, the minimum water need for bathing, washing clothes and swimming was considered to be 30 litres per capita per day (lpcd). Although this is less than the basic survival need of 50 lpcd identified by Gleick (1998),30 lpcd is reasonable as the HH needs for drinking, cooking, cleaning utensils and sanitation are met from other village sources.

The value of the domestic uses (washing and bathing) and livestock drinking was calculated by first estimating the cost which the water supply agency has to incur to produce and supply water using the technologies available in the area and then

Table 8.8 Quantum of catch of different varieties of fish from the five selected tanks, and the market value of the catch

Sr. no.	Name of tank	Number of families involved in fishing	Average amount of fish caught by the local communities, by variety (kg)							Total value of the fish in the market (INR)[a]						
			Rohi (Indian major carp)	Bhakur	Mirkali	Balia	Magur (catfish)	Grass carp	Small fish	Rohi (Indian major carp)	Bhakur	Mirkali	Balia	Magur (catfish)	Grass carp	Small fish
1	Gadloisingh	Fish contractor	80	70	120	40	0	0	0	5600	9600	3200	0	0	0	0
2	Jhankarpalli	Do	40	25	35	12		15	15	3200	2000	2800	960		1200	750
3	Laida	Do	45	40	35	20	18	15	20	3600	3200	2800	1600	1440	1200	1000
4	Rengloi	Do	60	35	40	25	15	35	30	4800	2800	3200	2000	1200	2800	1500
5	Rugudipali	Do	30	25	20	10	0	0	0	2000	1600	800	0	0	0	0

Source: authors' own analysis of primary data.

Note

a. The average market rate is taken as INR 80 per kilogram for all types of fish except for small fish (INR 50 per kg).

Table 8.9 The extent of the use of tank water for domestic and recreational needs

Sr. no.	Name of tank	Number of families using tank for (X)				Number of months for which tank is used for (Y)			
		Washing clothes	Bathing	Swimming	Livestock drinking	Washing	Bathing	Swimming	Livestock drinking
1	Gadloisingh	55	55	55	46	12	12	12	12
2	Jhankarpalli	180	180	180	180	12	12	12	12
3	Laida	500	500	500	413	12	12	12	12
4	Rengloi	134	134	134	8	12	12	12	12
5	Rugudipali	22	22	22	50	12	12	12	12

Source: authors' own analysis of primary data.

multiplying it with the minimum volume of water required to meet these needs in a year (the public pricing method) and the total number of person years/livestock years.

The basis of the estimates of the cost of water supply is as follows. Bore wells are used as a decentralized drinking water supply source in western Odisha. The cost per cubic metre of water that a public utility has to incur depends on the depth to water table and the aquifer characteristics. A higher depth to water table means a higher unit (variable) cost of pumped water. The capital cost also depends on the depth to water table and the aquifer characteristics. In the case of high yielding aquifers, the cost per unit volume of water would be lower, provided the depth of well does not change. We have considered a bore well in the region in the depth range of 180–200 feet, supplying water for domestic purposes. The cost of the system was considered to be INR 50,000 and INR 55,000 per unit, respectively. The cost per cubic metre (cost per m^3) of water was calculated on the assumption that the well would yield in the range of 1.5–2 litres per second. The life of the tube well was assumed to be 10 years and the discount rate was assumed to be 6 per cent. The annualized capital cost came to around INR 6,793 per system. The O&M cost was calculated to be INR 13,140 to INR 17,520 per annum (for running the pump for eight hours per day, for 365 days, with a 1.5 HP and a 2 HP pump respectively). The cost of water ranged from INR 1.39 per m^3 to INR 1.28 per m^3 of water (see Table 8.10). We have considered the mean of the two for our calculations. The cost does not include the cost of conveyance of the water through pipelines. This is done for the purpose of comparison, as just as in the case of tanks, the households will have to fetch the water from the source. It also does not include the cost of the operator: as in the case of tanks, the communities have to put in labour to fetch water from the source.

The estimated value of the social goods and services provided by the tanks are furnished in Table 8.11.

The total economic value generated by the tanks

The estimate of the total economic value generated by the tank considered the value of the economic output resulting from the use of water in irrigated crop production, in fish production (direct use values) and the economic value of social good produced by the use of the water for domestic purposes and livestock drinking. We have also examined the indirect use value of the water remaining in the tank through the recharge to the groundwater system. However, for the sandy clay loam soils, the benefit of recharge through infiltration is likely to be extremely low, with a steady state infiltration rate (under saturated conditions) of only 0.38 mm per hour (source: Texas Council of Government, as cited by Oram, 2009). Moreover, the soils of the wetland will, ideally, have a greater percentage of clay and silt, which further reduces the infiltration capacity of the tank bed. Hence, infiltration, which would occur only during the first few hours of the rains, can be considered negligible in the case of the wetlands in Sambalpur. Thus the recharge benefit was not considered. The estimates of the economic value of benefits derived from various tank (direct) uses are summarized in Table 8.12.

Table 8.10 Estimated public cost of the water supply in the villages in Sambalpur: two scenarios

Sr. no.	*Well discharge (litre per second)*	*Total volume of water pumped (m^3)*	*Capital cost of the system (INR)*	*Life of the system (years)*	*Annualized capital cost (INR)*	*Pump capacity (HP)*	*Annual O&M cost (INR)*	*Annual maintenance cost (INR)*	*Cost per m^3 of water*
1	1.5	15768	50,000	10	21933	1.5	13140	2000.00	1.39
2	2.0	21024	55,000	10	26945	2.0	17520	2000.00	1.28

Source: authors' own estimates based on secondary data.

Table 8.11 Value of the social good and recreational service provided by the tanks

Sr. no.	*Name of tank*	*Number of person years*[a]*/Number of livestock years*[b]				*Value of the service in a year*	
		Washing clothes	*Bathing*	*Swimming*	*Livestock drinking*[b]	*Washing, bathing*[c]	*Livestock drinking*
1	Gadloisingh	176	176	*176*	92	3974.85	2216.3
2	Jhankarpalli	226	226	226	360	13008.6	8672.4
3	Laida	256	256	256	826	36135.0	19898.3
4	Rengloi	195	195	195	16	9684.18	385.4
5	Rugudipali	101	101	101	100	1589.94	2409.0

Source: authors' own analysis of primary data.

Notes

a. The value for this is estimated by multiplying the average number of family members, and the number of months for which the source is used in a year (n) and dividing it by 12 (total number of months in a year), i.e., X*Y*n/12.

b. In the case of livestock, we have assumed that those families who use tank water for livestock hold an average of two animals (cows or buffalos) per household, and that the total water required per animal would be 50litres per day.

c. For estimating the value of the recreational service provided by the tank (swimming), hedonic pricing can be employed. However, since the area has a large number of water bodies, with each village having several of them, the value of this was considered insignificant.

Table 8.12 The economic value of various uses of water during normal and drought years

Sr. no.	*Tank name*	*Rainfall year*	*Annual economic value of the use of water (INR) in*				
			Irrigation	*Fisheries (by contractor)*	*Domestic use*	*Livestock use*	*Total economic value (INR)*
1	Gadloisingh	Normal	270374.00	18400.00	3974.85	2216.30	294965.2
		Drought	409403.00	18400.00	3974.85	2216.30	433994.2
2	Jhankarpalli	Normal	311594.00	10910.00	13008.6	8672.40	344185.0
		Drought	291346.80	10910.00	13008.6	8672.40	323937.8
3	Laida	Normal	269958.00	14840.00	36135.0	19898.30	340831.3
		Drought	252239.80	14840.00	36135.0	19898.30	323113.1
4	Rengloi	Normal	625069.60	18300.00	9684.18	385.40	653439.2
		Drought	195074.50	18300.00	9684.18	385.40	223444.1
5	Rugudipali	Normal	2152607.00	4400.00	1589.94	2409.00	2161006.0
		Drought	287973.00	4400.00	1589.94	2409.00	296371.9

Source: authors' own analysis using primary data.

The economic value generated from various existing uses of tanks is highest for tank 5, i.e., the Rugudipali tank, followed by the Rengoli tank. The high value in the case of the Rugudipali tank in the normal rainfall year is because of the large reported area of irrigation for that tank during kharif season. This figure may require thorough scrutiny. Nevertheless, the estimates of economic value are much higher for normal rainfall years in the case of tanks 2 and 4, the tanks wherein irrigated winter crops are produced. The analysis also shows that in none of the tanks is the value of the economic output generated from fishery activity higher than that from irrigated crop production.

Water allocation rules and technological changes for future tank management

Once the social needs of water are taken care of (domestic water use and livestock uses), water productivity, in economic terms, forms the basis for water allocation, in that it means maximizing the surplus value product from the uses of water in the tank. This is clear from Equation 3 (see above, under 'Analytical Procedure for Wetland Valuation').

The estimated values of applied water productivity in different kharif and winter crops grown in the tank command, and the volume of water allocated to those crops in typical rainfall years (i.e., a normal year and a drought year), are given in Table 8.13. The results clearly shows that water productivity, in economic terms, from irrigated vegetables such as brinjal, onion and tomato, and cash crops, such as mustard, grown during the winter season, are much higher than that of kharif paddy, which receives supplementary irrigation during the monsoon season. Further, the same crop gives higher water productivity during normal rainfall years when compared to drought years.

Table 8.13 The water productivity of different crops during normal and drought years

Tank	*Type of year*	*Paddy*		*Brinjal*		*Mustard*		*Onion*		*Potato*		*Tomato*		*Others*	
		INR/ m^3	*Volume (m^3)*	*INR/ m^3*	*Volume (m^3)*	*INR/ m^3*	*Volume (m^3)*	*INR/ m^3*	*Volume (m^3)*	*INR/ m^3*	*Volume (m^3)*	*INR/ m^3*	*Volume (m^3)*	*INR/ m^3*	*Volume (m^3)*
Gadloisingh	Normal year	25.12	5518.80												
	Drought year	22.93	17108.5												
Jhankarpalli	Normal year			101.38	1723.80			27.62	3864.20	7.13	1646.80	59.88	4154.80	27.62	1204.20
	Drought year	2.10	5607.60	34.81	2140.10			23.26	2285.60	4.63	1045.90	56.69	1573.90	17.49	837.70
Laida	Normal year	[See note[a]]													
	Drought year	8.22	16774.70												
Rengloi	Normal year	9.0	5348.20			273.23	3220.90			NA	1612.0			55.64	3233.30
	Drought year	12.24				288.29	563.60				411.0			20.47	1346.10
Rugudipali	Normal year	404.92	5211.60												
	Drought year	22.19	14968.80												

Source: authors' analysis using primary data.

Notes

a. The applied water productivity value is not estimated as the value of the denominator, i.e., the irrigation dosage is zero.

Water allocation to a cropping system which is dominated by vegetables gives higher overall net returns per unit volume of water and generates a higher surplus value product from irrigation. This is found only in the case of two villages. In the other villages, there is no cultivation of vegetables, and water allocation is inefficient. This is particularly true in view of the fact that earmarking more water during normal rainfall years for fishery does not lead to increased fish production (see the explanation in the conceptual framework on water allocation).

The results show that water allocation for agriculture during normal rainfall years is less than that of drought years in three out of the five tanks, in spite of having more water available during such years. This is partly because of the much lower water requirement for paddy during kharif in normal rainfall years, and partly because of the restriction on water release for irrigation during winter, as found in tanks 1, 3 and 5. In the remaining two tanks, where the volumetric water use for irrigation is greater during normal years, increased water allocation for irrigation was possible only because of winter irrigation of mustard and vegetables.

Judging by Equation 3, at present there is a sufficient scope for improving the economic value of tank water used for irrigated agriculture in three ways: a) reallocating the water used for growing kharif paddy to winter crops during the drought years, as the water required for paddy is very high during these years; b) using some more water from the tank for crop production during the winter season, in normal rainfall years; and c) increasing the utilization potential of the tank water during the kharif season of normal rainfall years, by taking it out of the tank command using conveyance systems. The crops that can be grown are brinjal, tomato, potato, onion, mustard and some curry leaves, such as fenugreek and coriander. This would enhance the economic outputs from the tanks remarkably. But, while doing this, their impact on domestic food security needs to be thoroughly examined.

Findings and conclusions

Analysis of five wetlands in western Odisha shows that that there are five major uses of water from the wetland. They are domestic water use; livestock drinking; irrigation; fish production; and recreation (swimming). The economic value of the multiple use benefits created by the direct use of water from the tank was estimated by several methods, viz., the hedonic pricing method, market analysis and the public pricing method. The total economic value (TEV) of various uses ranged from INR 2.95 lac in a normal year in the Gadloisingh tank to INR 21.61 lac in the case of the Rugudipali tank in a normal year. Interestingly, the incremental return from the use of water per unit area of irrigated crop in the wetland over upland was found to be higher during drought years for three tanks, indicating the greater value the tank water has for the farmers in such years. Nevertheless, at the aggregate level, the incremental return from the use of tank water was found to be lower in four out of the five tanks, indicating the distress of poor tank irrigators. Irrigation produces the highest value in economic terms in all the tanks.

Water allocation from the tanks for agriculture during normal rainfall years is less than that of drought years in three out of the five tanks, in spite of having more water available during such years. In the remaining two tanks, where the volumetric water use is greater during normal years, this was made possible through winter irrigation. The extremely low water requirement for paddy grown during the kharif season, the inability to expand the command of the tank, and the restriction on water withdrawal during the winter season (imposed by the Panchayat, which leases out the tank to contractors for fish production), are the reasons for this. Because of this restricted water allocation, the economic value of the benefits realized from the use of tank water is quite low during normal years. Irrigated agriculture is in direct conflict with fishery production.

Water productivity analysis shows that all of the winter crops, except potato, have higher water productivity, as compared to kharif paddy, during normal as well as drought years. Also, the same crop, grown in the wetland, yields higher water productivity during normal years as compared to drought years. A conceptual model was developed to analyze the gross tank product from the direct uses of tank water for different water availability scenarios. Subsequently, mathematical formulations were derived for simulating the economic value of various benefits derived from the tank under various water allocation scenarios, with various physical and socio-economic constraints induced as constraints and boundary conditions.

The application of this model shows that there are no significant trade-offs between maximizing the economic value of water in agriculture production, and meeting water needs for other existing uses of the tanks. The options for improving the economic value of tank water used for irrigated agriculture, without compromising on basic needs and fisheries, are: a) reallocating water used for growing kharif paddy to winter crops during the drought years, as the water required for paddy is very high during these years; b) using some more water from the tank for production of crops that are high valued, and that which give high economic returns per unit volume of water during the winter season, in normal rainfall years; and c) increasing the utilization potential of the tank water in normal rainfall and wet years, in the kharif season itself, by taking it to areas outside the command through new conveyance systems.

But, to affect the enforcement of these rules, strong institutional intervention is required. Water reallocation is the biggest challenge here. First, there should be sufficient infrastructure to expand the command area of the tank during normal rainfall years. Second, there should be an institutional mechanism to ensure that sufficient water from the tank is earmarked for winter crop production. This can be done without compromising on fish production, which needs the presence of a minimum quantity of water for longer time periods. This should be supported by good scientific and technical knowledge of growing horticultural crops and raising fish. Periodic water quality testing is required to ensure good quality water for domestic use and fisheries. It is to be kept in mind that keeping a lot of water in the tank for fisheries, without allowing it to be used for crop production, does not result in increased fish yield.

Multiple use water systems for developing economies

Conventionally, rural water systems are designed as single use systems. But communities in rural areas require water for both domestic and productive needs. Water supply systems that are not responsive to the livelihood needs of rural communities fail to play a much needed role; yet the communities also show low levels of willingness to pay for the water supply services. This would affect the sustainability of the systems as official agencies will not be able to recover the costs of their operation and maintenance (GSDA, IRAP and UNICEF, 2011). In reality, when such systems are built, by default they become multiple-use systems. This is particularly the case in developing countries, where water for basic survival and livelihood needs is still an issue (van Koppen et al., 2009). It is common to find rural drinking water sources also being used for washing and feeding dairy animals and watering kitchen gardens, as well as catering to domestic uses. Such unplanned uses often damage the system (Moriarty et al., 2004; GSDA, IRAP and UNICEF, 2011). There is growing recognition of the fact that much social and economic value can be realized from the diversion of water from single use systems to multiple needs, without inflicting damage to the infrastructure (van Koppen et al., 2009; Moriarty et al., 2004).

The past decade has seen many experiments on multiple use water systems in developing countries such as India, Nepal, Bangladesh, Kenya, Ethiopia, Thailand and Sri Lanka. They mostly concern the use of water from large irrigation systems for uses other than irrigation, such as domestic uses, fisheries, livestock drinking, recreational uses, micro hydro power generation and small industrial units in rural areas (source: based on GADA, IRAP and UNICEF, 2011). Though water uses compete, conflicts over water sharing do not exist in these systems as water is available in plenty. Very little scientific literature concerns the use of small water bodies in rural settings, such as tanks and ponds, for multiple purposes, particularly their functioning when subject to high variability in water supplies due to climate. In such situations, since there could be conflicts over the use of water from these systems, water allocation forms an important issue.

Unlike in the case of the Odisha tanks, there could be trade-off between maximizing the economic value of water from the tanks (gross tank product) and realizing the social objective of meeting multiple water needs. Mechanisms have to be established for compensating those users who are likely to suffer losses due to such reallocation decisions, in order to resolve future conflicts. In certain cases, technological solutions might come to the rescue of water users. For instance, as seen in Andhra Pradesh in southern India, wherever topography permits, water from large irrigation schemes is released into minor irrigation tanks and drinking water tanks through the existing canal networks and feeder channels specially constructed for the purpose. This system helps significantly to minimize the crisis during years of droughts. Conflicts over water sharing would be fewer when there is sufficient water in the tanks.

In most of sub-Saharan Africa, and some countries of South Asia, the development of rural water infrastructure for drinking water supplies or irrigation is very poor. The most water-stressed countries of east, central and west Africa, which have the poorest water supply infrastructure, need priority attention. There is enormous scope for applying the concept of multiple-use systems in such situations. The schemes for provision of water in rural areas have to be designed for multiple uses. But there are several issues which need to be addressed. If the systems, which are predominantly for irrigation, have to serve domestic water needs, the issues of quality and access will have to be addressed. Since irrigation infrastructure is built to deliver water to the farms, there will be a need for providing additional infrastructure to take water to the habitations. For rural water supply schemes to be redesigned as multiple-use systems, the quantity issue needs to be addressed. This is because the amount of water required by communities for productive needs (such as livestock drinking, the kitchen garden and small rural industries) would be far higher than that required to meet human needs.

Generally, if village water supply schemes have to cater to multiple needs, the dependability of the source needs to be ensured for sustainability of the system. As sub-Saharan Africa is prone to drought and experiences high variability in rainfall (Foster et al., 2006), this issue is particularly important for that region. Under such circumstances, local water sources, surface and underground, can deplete during years of poor rainfall. The available water will not be sufficient even to meet basic survival needs, threatening the sustainability of the system. Therefore, at the design stage, the scope for integrating the local sources with more reliable regional schemes have to be explored, and infrastructure built.

Notes

1 An open well of 50 feet in depth and with a diameter of 10 feet would cost around INR 70,000 to INR 75,000, including lining with granite stones or bricks.

2 Here, it is to be kept in mind that some of the water would be lost in evaporation while storing it for winter and summer use, and the water available in the tank would be up to the level indicated by "NL" in the diagram. This is the same as the water which is available for fisheries.

References

Asian Development Bank. (2006). *Rehabilitation and management of tanks in India: A study of selected states*. Manila, Philippines: Asian Development Bank.

Costanza, R., Wainger, L., Falke, C., & Maler, K-G. (1993). Modeling complex ecological economic systems. *Bioscience, 43*, 545–55.

Foster, S., Tuinh of, A., & Garduño, H. (2006). *Groundwater development in sub-Saharan Africa: A strategic overview of key issues and major needs*. Case profile collection 15. Washingon, DC, USA: The World Bank.

Gleick, P. H. (1998). Water in crisis: Paths to sustainable water use. *Ecological Applications, 8* (3), 571–579.

Government of Orissa. (2007). *Report of the high level technical committee to study various aspects of water usage for Hirakud Reservoir*. Bhubaneswar, India: Water Resources Department, Government of Orissa.

Groundwater Survey and Development Agency/Institute for Resource Analysis and Policy and UNICEF. (2011). *Multiple use water services to reduce poverty and vulnerability to climate variability and change*. Feasibility report of a collaborative action research project in Maharashtra, India. Mumbai, India: UNICEF.

Kumar, M. D. (2007). *Groundwater management in India: Physical, institutional and policy alternatives*. New Delhi, India: Sage Publications.

Kumar, M. D., Narayanamoorthy, A., & Sivamohan, M. V. K. (2009). *Irrigation water management for food security in India: The forgotten reality*. Paper presented at the National Seminar on Food Security in India, Greenpeace Foundation, New Delhi.

Moriarty, P., Butterworth, J., & Van Koppen, B. (Eds.). (2004). *Beyond domestic: Case studies on poverty and productive uses of water at the household level*. IRC Technical Paper no. 41. Delft, The Netherlands: IRC.

Oram, B. (2009). *Soils, infiltration and on-site testing*. Wilkes-Barre, Philadephia, USA: Wilkes University Geo-Environmental Sciences and Environmental Engineering Department. http://www.water-research.net.

Palanisami, K., Meinzen Dick, R, & Giordano, M. (2010). Climate change and water supplies: Options for sustaining tank irrigation potential in India. Review of Agriculture. *Economic and Political Weekly*, *45* (26 & 27), 183–190.

Sahu, G. C., & Nanda, S. K. (undated). *Studies on the clay mineralogy of Orissa soils*. Bhubaneswar, India: Orissa University of Agriculture and Technology Department of Soils and Agricultural Chemistry,

Shah, T. (2001). *Wells and welfare in Ganga basin: Public policy and private initiative in eastern Uttar Pradesh, India*. Research Report 54. Colombo, Sri Lanka: International Water Management Institute.

Sikka, A. K. (2009). Water productivity of different agricultural systems. In M. D. Kumar, & U. Amarasinghe (Eds.), *Water productivity improvements in Indian agriculture: Potentials, constraints and prospects*. Strategic analyses of the national river linking project (NRLP) of India, series 4. Colombo, Sri Lanka: International Water Management Institute.

Turner, R. K., van Den Bergh, J. C. J. M., Barendregt, A., & Maltby, E. (1998). *Ecological-economic analysis of wetlands: Science and social science integration*. In T. Soderqvist (Ed.), *Wetlands: Landscape and institutional perspectives* (pp. 327–350). Beijer occasional paper series. Stockholm, Sweden: Beijer International Institute of Ecological Economics.

van Koppen, B., Smits, S., Moriarty, R., Penning, F. Devries., Mikhail, M., & Boelee, E. (2009). *Climbing the water ladder: Multiple use water services for poverty reduction*. TP series 52. The Hague, The Netherlands: IRC International Water and Sanitation Centre and the International Water Management Institute.

9 Future strategies for agricultural growth in India

M. Dinesh Kumar, V. Niranjan, A. Narayanamoorthy and Nitin Bassi

Introduction

While the government continues to make huge investments in irrigation-related infrastructure, of late there has been an over-emphasis on the augmentation of groundwater resources and the conservation of water in small water bodies, with a consequent focus on decentralized water harvesting and local groundwater recharge, in schemes such as NREGA (Bassi and Kumar, 2010). But there are no long-term strategies in sight for supporting large-scale irrigation development or promoting productivity in agriculture.

While small water harvesting structures would no doubt provide reliable local water supplies in high rainfall areas, in most parts of India this would not make much hydrological sense, nor be a sound economic proposition (Kumar et al., 2008a). As a result, many negative effects are observed, particularly on the agricultural front, distorting labour markets (Bassi and Kumar, 2010). Added to this, the policies governing the use of water in agriculture are degenerative, driven by political considerations, and promote inequity in accessibility and the inefficient use of water. They defeat the very goal of sustainable agricultural production. In this chapter, we examine the effectiveness of the government's current policies and programmes for agricultural growth. In order to do this, an analysis of the performance of the agriculture sector in the 11th Plan period and scientific evidence available from recent research on the topic is undertaken.

New strategies for agricultural growth and food security

The 11th Plan strategy of inclusive growth rests upon a substantial increase in the plan allocation for agriculture, and irrigation and water management. While the allocation for agriculture and the allied sectors (at 2006–07 prices) is INR 54,801 crore (US$ 1 equates to approxmately INR 50), for irrigation and water management it was INR 3,246 crore. Since irrigation is a state subject, there is a major contribution from state plan allocations, to the tune of INR 1,82,050 crore. In addition, the contribution from schemes such as the Accelerated Irrigation Benefit Programme stands at INR 47,015 crore (GOI, 2008, pp. 42–62). Hence, the total plan allocation for irrigation was INR 2,32,311 crore. This is far higher than the allocations made during the 10th Plan period for irrigation and flood control, which was only INR 84,692 crore. Similarly, in agriculture and allied sectors, the plan expenditure was only INR 19,175 crore (GOI, 2008).

But, in spite of the substantial increase in plan allocation for agriculture and irrigation, agricultural growth rates have not shown any encouraging trend. While the growth rates in sectoral GDP was 2.47 per cent during the 10th Plan period (GOI, 2008, Table 1.1, p. 4), during the first two years of 11th Plan period, the average growth rate was only 3.0 per cent. This was in spite of the fact that there had been a threefold increase in plan allocation. There are two important reasons for this. First, while irrigation is the key to boosting agricultural growth in India, the overall progress in public (surface) irrigation development is poor, owing to cost and time over runs in project completion, which results in a very high capital cost of bringing one hectare of cultivable area under irrigation. Second, while some additional irrigation is achieved through the new schemes, there is a gradual decline in the performance of old schemes due to dwindling reservoir capacity and lack of proper upkeep of water distribution and delivery infrastructure, which offset the gains. As pointed out in the second chapter of this book, the other reasons could be a) increased diversion of water from surface reservoirs for urban domestic uses; and b) the reduced flows into reservoirs due to intensive water harvesting and watershed development in the upper catchment areas.

In addition to the expenditure under agriculture, irrigation and flood control, there has been substantial investment under the National Rural Employment Guarantee Scheme (NREGS), ever since 2007–08, when it was first implemented in all districts of the country. On average, the government allocates around US$ 9 billion (INR 40,000 to INR 45,000 crore) every year for this scheme. The activities being undertaken under NREGA are drought-proofing (afforestation and tree plantation); water conservation and water harvesting; construction and cleaning of minor irrigation canals; renovation of water bodies; flood control and protection works including embankments; land development; and the creation of irrigation facilities for poor SC/ST (scheduled caste and scheduled tribe) households. Though started as a social welfare scheme aimed at rural employment generation, with many components purely on water management, it is promoted as a strategic intervention for agricultural growth in rain-fed areas.

Barring construction of roads, a large proportion of the infrastructure created under NREGS directly affects the rural water sector (Bassi and Kumar, 2010), and therefore agriculture and allied activities. However, the planning of this work, which is done in a decentralized format, does not take into consideration the hydrology, geo-hydrology, topography and agro-climate of the localities concerned (NCAER-PIF, 2009), nor the cost effectiveness (Kumar and Bassi, 2011). A major proportion of the funds were spent in road building in areas which are most suitable for watershed development (Ambasta et al., 2008). Recharge structures are built in hard rock areas without considering the geological and geo-hydrological features, which heavily influence the performance of such structures (Bassi and Kumar, 2010). Tree plantation is taken up in semi-arid and arid low rainfall areas (Tiwari et al., 2011) without much attention being paid to the availability of water for providing protective irrigation. The cost per cubic metre of water for these structures is not worked out and compared against that for other options.

Analysis shows that the NREGA interventions cause more negative welfare effects than positive ones, the most important among them being an artificial scarcity of agricultural wage labour and an unprecedented increase in wage rates (Bassi and Kumar, 2010; Panagaria, 2009). As noted by Bassi and Kumar (2010), in naturally water-scarce basins, indiscriminate construction of water conservation structures such as ponds and check dams, and the de-silting of existing village water bodies, is causing negative impacts on the overall water economy. This is as a result of reduced inflows into downstream reservoirs meant for irrigation, domestic and industrial purposes, and cost in effective water harvesting structures. There is a low annual rainfall and high inter-annual variability which causes disproportionately higher variability in runoff, high potential evaporation, low infiltration rates and poor groundwater storage potential in these regions (Kumar et al., 2008a). The land-based NREGA interventions lack both proper scientific planning based on hydrological considerations and technical supervision of the work execution, impacting on their overall effectiveness (Bassi and Kumar, 2010).

But the real source of the agricultural growth in recent years is not food grain crops, but the high valued fruits, vegetables, milk and fish. Over the years, the contribution of these high valued crops to agricultural output, in value terms, has been increasing alongside the increase in the production of fruits, vegetables, milk and fish (Nandakumar et al., 2010). In fact, the per capita availability of food grain over the triennium has been on the decline since 1990. It declined from 172.5 kg per year in 1990 to 159.2 kg per year in 2008 (provisional). Against this, the per capita availability of milk, egg, fish, fruits and vegetables has been increasing over the years. Some scholars argue that the food security issue is not so much about the availability of food grains, but about its economic access. But subsequent analysis would show that this is not true. For instance, one of the arguments put forth by scholars to discard the food availability argument is the changing consumption pattern, with an increasing preference for animal products (chicken, egg and milk), fruits and vegetables (see for instance, Nandakumar et al., 2010). While it is expected that such changes in the consumption pattern would reduce the pressure on cereals for direct consumption, it would prove to be costly if one was to ignore the fact that even increased production of egg, milk, poultry and inland fish would require a large amount of cereals in the form of animal and poultry feed (Amarasinghe et al., 2007). This would place an additional burden on cereals, the production of which is already threatened. This would accelerate the rise in the price of cereals, and mainly impact the poorer sections of society.

The price of food grains, including cereals and pulses, has been on the rise recently, with food inflation touching double digits (Chowdhury, 2011). The food security of hundreds of millions of poor people will be at risk if we do not increase the per capita production of cereals in proportion to the growing population.

Sustainable agricultural production and food security

The ideal future strategy should include both improvement in productivity in the use of water, which, after all, is a scarce resource, and an effective increase in the availability of utilizable water for irrigation. But water abundant regions have very little arable land that can be brought under irrigated production, and are dependent on food imported from land-rich, water-scarce regions, which maintain high levels of productivity.

Hence, the future of irrigation development in India lies in the appropriation of surface water in the water-abundant basins, and its export to, and use in, those water-scarce river basins that are endowed with sufficient amount of arable land. This would help boost agricultural production, along with improving the sustainability of groundwater use. But this will not be possible everywhere, due to topographical and engineering limitations. In water-scarce regions, the emphasis should be placed on improving water productivity in agriculture. Hence, the focus should be on economic instruments, such as water and energy pricing. But too much reliance on this will also have its problems, such as excessive preference for high valued crops that use less water, with long-term negative consequences for cereal production. In the long run, large-scale water imports would be required in the semi-arid and arid regions which are agriculturally prosperous, in order to save these regions from drought and sustain irrigated agriculture. In water-abundant regions, the policies and programmes should be designed to encourage more intensive use of water.

Technologies to change the trajectory of irrigation development

Manually operated pumps and micro diesel engines in water-abundant regions

Several regions have abundant groundwater supplies, including Assam, parts of Bihar, Odisha, West Bengal and Jharkhand (GOI, 2008). These regions accommodate the largest number of poor people in the country (Shah et al., 2000). The groundwater resources in these regions largely remain under-utilized in spite of the fact that public irrigation facilities are very poor (GOI, 2008). As a result of the conventional (energized) water abstraction structures and mechanisms prevalent, the trajectory of development of groundwater resources in the region is likely to be very low.[1] In order to spur the development process, these regions need simple technologies which involve very little capital investment, and which can absorb the surplus labour force. The region can thus promote equity in access to groundwater for irrigation, while increasing its utilization. For the poor small and marginal farmers, who are exploited by the rich for irrigation water (Kumar, 2007), this would provide relief. Though expansion in irrigation through the manual pump is not expected to be significant, the remarkable achievement would be in the provision of water security for millions of marginalized farmers.

The treadle pump (TP), a manually operated pump, requires very low capital investment, while being much more energy efficient than traditional water lifting devices, such as denkul and shena. The pump, which costs in the range of INR

1,000 to INR 1,400, is highly suitable for the millions of poor farmers in the region who have tiny holdings. It can provide them the water security that is essential for their livelihoods. In eastern India, adoption of treadle pumps led to an expansion in the irrigated area, cropping intensities, enhanced crop output and yield, and a significant rise in income through farming, while farmers move from subsistence agriculture to wealth-creating irrigated farming practices (Shah et al., 2000; Kumar, 2000b).

Studies conducted in Odisha also provide enough hard empirical evidence to show that pump adopter households enjoy greater food and nutritional security (see Kumar, 2001, for details). Treadle pump irrigation ensured increased output from irrigated crops and, more importantly, vegetables. The surplus production sold in the market provides cash income from which other essential commodities can be purchased (Kumar, 2000b). Another aspect of household food security is the nutritional value of the food consumed. The study found that the introduction of TP has directly contributed to a growth in vegetable production from farms.

In addition to treadle pumps, effective rural electrification, easy access of farmers to farm power connections, and the availability of credit facilities/subsidy for purchase of pump sets would improve the access of small and marginal farmers to groundwater irrigation through two different routes. First, an increase in the number of pump owners would reduce the monopoly power of pump owners, who otherwise charge exorbitant hourly rates for irrigation (Kumar, 2007). Second, instead of purchasing water, more farmers would opt for power connections and electrification of their shallow wells, making groundwater irrigation cheap for a larger constituency of farmers.

Recently, some argument have been made that three factors have created conditions unfavourable for the increased adoption of treadle pumps. Two of them are increased employment opportunities in rural areas, especially after the launch of NREGA and the subsequent increase in wage rates for labourers; and the increased availability of Chinese micro diesel engines in eastern Indian states. While several treadle pump owners have shifted to diesel engines in the recent past, it is going to be a temporary phenomenon. In fact, increasing fuel prices are likely to once again attract the small and marginal farmers towards treadle pumps. Another reason to opt for treadle pumps is related to the rising price of vegetables and the large increase in the number of small urban centres in the country. Research amply shows that, even in the traditional paddy growing areas, the treadle pump owners grow high valued vegetables and root crops and transport and sell them in the nearest towns, maximizing the returns from the labour and other inputs they use. The foregoing analyses suggest that treadle pump irrigation would be highly profitable for households which have very small holdings and surplus labour.

Technologies for water productivity improvement

So far as harnessing more and more water from the natural systems is concerned, the technologies described above have their limitations. The other option available to enhance food production is to improve the efficiency of the use of water (Kijne et al.,

2003; Kumar, 2010). Worldwide, micro-irrigation (MI) technologies are promoted to save water and enhance the increased efficiency of water use in agriculture (Postel, 1992; Postel et al., 2001). There are several technologies which help farmers save irrigation water (Kumar et al., 2008b). While MI systems have seen a relatively rapid adoption rate over the past decade in India, the overall adoption level is still quite low. Drip and sprinkler irrigation systems cover less than 6 per cent of the global irrigated area and, in the case of India, they cover 3.88 milion ha, with 1.46 million ha under drips and 2.42 million ha under sprinklers.

But, the advantages of the technologies are biased in favour of farmers with large holdings. To enable farmers take full advantage of the water-saving potential of the technologies, they should install drip and sprinkler systems in large fields. However, in areas (states) where power pricing is dependent on the pump's horse-power, both the capital cost of the pumping per unit area and the operating cost per unit area will be higher for resource-poor farmers who adopt the system for smaller areas. Those farmers who adopt the system for larger areas can reduce the cost per unit area significantly (Kumar, 2003; Kumar, 2009). These systems involve high capital investments. Furthermore, the maintenance requirements for these irrigation systems are very high. The drip system, which is the most water-efficient of these technologies, is most suitable for horticultural plantations, in terms of cost effectiveness. Thus, they are best suited to resource-rich, large farmers, who can spare part of their land for horticultural crops and can wait for 3–4 years for returns. Another important issue involved in the adoption of pressurized irrigation systems is the lack of economic incentives. In many Indian states in which depletion problems are encountered (such as the alluvial areas of north Gujarat, Punjab, western UP and Haryana), groundwater resources are abundant: only power supply is limiting the farmers' access to groundwater. In these situations, groundwater supply potential is higher than that which the available power supply can deliver (Kumar et al., 2008b; Kumar, 2009).

The large static storage of the aquifers permits the farmers to keep pumping water, even though it causes excessive draw-down. This is because, first, either the cost of electricity for pumping per unit volume of water is extremely low, or the marginal cost of energy for pumping is negligible. Second, there are no limits on the volumetric pumping by well owners, and no payment is required for the water (Kumar, 2003, 2009). Since pressurised irrigation systems need extra power to run, the well output could drop with the installation of the system. As the farmer is already utilising the power supply fully, the total water output from the well would drop. Thus, the farmer would not be able to capitalize the benefit accruing from water saving in the form of an increased area under irrigation. Hence, the only economic opportunity available with pressurized irrigation technologies is yield increase. However, the ability to secure a higher yield through water saving devices depends heavily on management practices, including agronomic practices (Kumar, 2003).

The situation would be vastly different in hard rock areas facing depletion problems. Currently, farmers are not able to utilize the power supply fully, due to a shortage of water in the wells in these areas. In such situations, pressurized irrigation systems could benefit the farmers by enabling them to run the pump for longer hours, maintaining the same level of total well output and irrigating a larger area. Water-saving technologies to suit the requirements of many millions of the poor, small and marginal farmers do exist in the country. They are the mini sprinkler systems and micro tube drip systems (Kumar, 2003, 2009).

One important factor which limits the adoption of MI technologies is small sized fields (Kumar, 2009). MI systems are most suited to fruit crops, vegetables, tubers and flowers. The area under these crops is still very limited in India, with fruits accounting for 4.96 million ha, vegetables 6.76 million ha, and flowers 1.16 lac ha (GOI, 2008, p. 17). But the water saving impact of MI systems depend on a variety of factors, such as crops, type of MI technology, soil type, climate and geo-hydrology. Water saving is likely to be greater in the case of widely spaced crops under sandy soils in a semi-arid to arid climate where there is a deep groundwater table with drip or trickle irrigation (Kumar et al., 2008b; Kumar, 2009). It cannot be assumed that MI system adoption for any of these crops would result in a water-saving benefit, irrespective of the technology used and the soil type, climate and geo-hydrology of the area in which they are grown.

In accordance with this argument, the (real) water saving from MI systems should be relatively high in naturally water-scarce regions, and low in naturally water-rich regions (those characterized by high rainfall, humidity, and shallow groundwater table conditions). It has already been pointed out that naturally water-scarce regions also experience a physical scarcity of water, and, because of this correlation, the social benefit accrued from water saving would be very high in naturally water-scarce regions. As Kumar et al. (2008b) argue, these are the regions in which MI systems need to be subsidized.

The central funds allocated for subsidizing MI systems are very meagre in comparison to what is required to boost MI adoption in the country. A significant portion of the funds which are utilized for NGREA in naturally water-scarce regions, particularly for water conservation and drought-proofing activities, should be re-allocated to subsidize MI systems, so that they have a significant impact on water use efficiency in crop production and overall agricultural productivity.

Transfer of water from abundant to scarce regions

One of the new sources for the growth in aggregate demand for water in the country is the increasing need to meet ecological water demands in river basins. While agricultural water demand would remain the prime concern of future water demand management in India, this alone will not solve the growing water scarcity problems in many river basins. The reasons are many, the most important being the limited potential of MI technology. While MI technologies are best amenable to high valued fruits, vegetables and roots, the area under these crops cannot be expanded beyond a cetain limit, as to do so would curtail the area under food crops. The total (real) water saving possible through MI systems for five major crops that are amenable to MI systems would be 44.5 billion cubic metres (BCM) at the current cropping pattern (Kumar et al., 2008b), whereas the gap between water demand from various competitive use sectors and supplies would be 208 BCM. The scope for a major shift in cropping pattern to low water-intensive and highly water-efficient crops is extremely limited, due to food security concerns.

In order to meet the deficit, as argued in Chapter 3, the physical transfer of surface water from water-rich regions to water-scarce regions is crucial. There are two major gains from physical water transfer. There would a significant increase in utilizable freshwater resources, as the naturally water-rich regions of eastern India are short of arable land where the surplus water could be utilized for agricultural production. In Chapter 3, we have seen that the naturally water-rich regions are net importers of food from agriculturally prosperous water-scarce regions. An increase in the area under irrigated crop production in the water-importing regions would lead to greater agricultural surplus, increasing farmers' ability to export food to the water-rich regions (Kumar and Singh, 2005). Second, the economic value of water use in agriculture is higher in water-scarce regions than in water-rich. Hence, an increased value is realized through the physical transfer of water (Kumar et al., 2008c).

There are other benefits of physical water transfer for irrigation. As seen in the case of the Sardar Sarovar Project, gravity irrigation using imported water in the water-scarce regions leads to the increased recharge of groundwater through irrigation return flows (see Chapter 4 of this book). This reinforces our argument that this would improve the sustainability of well irrigation in those regions facing problems of groundwater over-draft. The reduced pressure on groundwater that results from there being an alternate source of water for irrigation would further improve the groundwater balance. A rise in water levels would also reduce the economic cost of the energy for groundwater pumping in irrigation. Such positive externalities of gravity irrigation would be enormous (for further discussion, see Chapter 4 of this book).

There are many regions in India which are agriculturally prosperous but facing an acute shortage of water for agricultural production. They are a) western India, comprising the western part of Rajasthan, north and central Gujarat, Saurashtra and Kachchh; b) the western part of central India, including parts of Madhya Pradesh and Maharashtra; and c) most parts of the southern Indian peninsula, covering almost the entirety of Andhra Pradesh except the coastal areas, most parts of Karnataka barring the western Ghat region, and almost the entirety of Tamil Nadu. They are located in water-scarce river basins.[2] The current water utilization scenario in the basins encompassing these regions shows that the future of agriculture there could be in great jeopardy if additional water resources are not provided to them. The main reason for this is that groundwater resources, which were the backbone of the agricultural economy in these regions, have been depleting alarmingly (Kumar, 2007).

Data from Andhra Pradesh indicate that well irrigation has started showing declining trends in the hard rock areas. The contribution of open wells, which used to be an important source of irrigation, to the state's irrigated area started declining in the early 1990s, with a steep drop after the late 1990s (Figure 9.1). On the whole, there has been hardly any increase in the net and gross area irrigated by wells in

Andhra Pradesh since 2004–05. Such an alarming trend can be seen in many hard rock areas in India. It appears that the only way to sustain well irrigation in these regions is through the importing of surface water, as the social consequences of groundwater depletion will put the livelihoods of millions of poor farm households at risk. Rural–urban migration and the increased indebtedness of farmers would be some of the immediate consequences. It is quite noteworthy that the regions with high rates of farmer suicides coincide with those having problems of groundwater mining and rampant well failures. Sustaining well irrigation is therefore essential in maintaining the social fabric of rural areas.

Transferring water to these regions from water-abundant basins, through inter-basin water transfers, requires complex engineering solutions. In addition, it is necessary to address complex social, economic, financial, legal and ecological questions involved in inter-regional, bulk water transfers. Ecologists and environmentalists have already raised alarm signals about possible "ecological disaster" in the donor basins in the wake of water transfers from the water-rich basins of eastern India (Ganges, Brahmaputra) to water-scarce basins in the west and the south. The other stakeholders of such interventions are the states which have to part with their water resources, and the issues involved here are of a political nature. Those political parties and governments in the states concerned try to make mileage out of the decisions, whether they have favourable or negative consequences for them. As the basins are inter-state and international, there are issues about the livelihoods of millions of people living in the lower riparian states and countries.

Theoretically, and also practically, what can hasten the decision to engage in water transfers is the fact that the opportunity cost of not undertaking such projects would be prohibitive. The social benefits which are likely to accrue from them, if executed, would result in the transformation of the country wide. But, from a political economy perspective, the pressure for large-scale water transfers

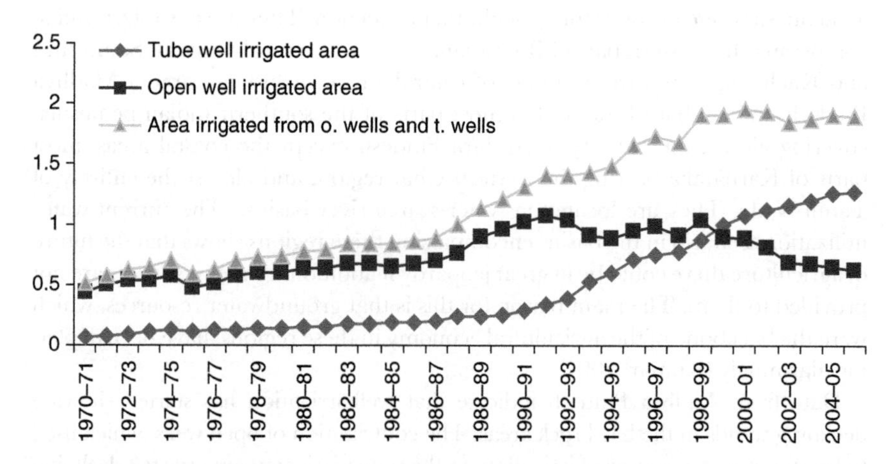

Figure 9.1 Net area irrigated by open wells, tube wells and total well irrigated area, AP.

Source: Directorate of Economics and Statistics, Government of Andhra Pradesh.

will spring from the urban areas. As detailed in Chapter 2, the metros located in naturally water-scarce regions are dependent on water imported from far off reservoirs to meet the lion's share of their water requirements. Today, the real urban growth is happening in the water-scarce western, north-western and southern regions, comprising cities like Delhi, Jaipur, Ahmedabad, Indore, Nagpur, Rajkot, Hyderabad, Pune, Bangalore, Vijayawada and Chennai, and in hundreds of other fast growing towns/centres. By virtue of being the engines of growth, the socially and politically influential urban areas are likely to put great pressure on the state and federal governments to provide water for their needs. The dire need to provide water to the prominent metros would ultimately help manage the politics involved in water transfer projects.

Once this becomes socially and politically feasible, the next challenge would be to ensure that water is stored and used optimally. While it is evident that importing water from large distance away is going to be a costly affair, making such projects economically viable requires that water is put to the most efficient use from an economic perspective. In other words, the productivity of use of water has to be very high. Water control would be the most critical element in determining the productivity of the precious water.

Groundwater banking and (intermediate) tank storage

Gravity irrigation has several limitations as a result of its inadequate control over water. Due to the poor reliability of the water supply and a lack of proper control over water delivery, canal commands generally favour water-intensive crops such as paddy, banana and sugarcane, which are not highly sensitive to excessive watering. But these crops have low water use efficiency in economic terms (INR per cubic metre). Therefore, alternatives to gravity irrigation are required when the cost of importing surface water for irrigation becomes prohibitively expensive. "Groundwater banking" is a viable option for effectively increasing water availability in areas where sufficient aquifer storage space is available (Contor, 2009; Hostetler, undated).[3] Groundwater banking here refers to the intentional infiltration of surface water and incidental recharge from irrigated crop land. This can become a good strategy in north-western and western India, as the over-exploited alluvial aquifers in these areas would provide storage space for the imported water. In the south, since the hard rock aquifers have limited storage potential, the infrastructure which transfers water in bulk from the eastern basins has to be integrated with hundreds of thousands of small and large water bodies spread all over the three states of the southern peninsula, namely Andhra Pradesh, Karnataka and Tamil Nadu.

Since conventional artificial recharge structures would be highly expensive, the cultivated land could be used as infiltration basins in the alluvial areas. This could be achieved by transferring the floodwater during monsoon to the land covered by the standing kharif crops, for incidental recharge.

When the recharge basin is large, with soils having good hydraulic conductivity, the recharging process would be effective during rainy season (Watt, 2008). Also, this would not involve additional costs. The natural recharge from rainfall in the region, plus the additional recharge from irrigation return flows, would give sufficient water for irrigating winter crops. Groundwater irrigation also encourages the use of efficient irrigation technologies to which gravity irrigation is not amenable. In the south, less water would need to be put into the hard rock aquifers through the mechanism explained above, as most water could be stored in the tanks. Since most of the tanks do not receive adequate natural inflows from their catchments, as indicated by their poor performance, storage space will not be a problem in most years.

Institutional changes for changing the trajectory of water use in agriculture

Promoting equity and productivity in water use

The growing competition and concomitant conflicts between different sectors are major issues that need to be addressed in water allocation. The fundamental challenges are the promotion of economically efficient uses, while adequately compensating the agriculturists for the losses they suffer due to transfer of water for other efficient use sectors; and equitable access to water from canals and groundwater within the agricultural sector (Kumar, 2010). Saleth and Dinar (1999) points out that concerns in the water sector, which once revolved around water development (and quantity), now revolve around water. Markets and regulatory approaches can be used as instruments for water allocation (Frederick, 1992), but both are likely to fall short of satisfying these criteria for efficient and effective water allocation (Frederick, 1992). The enormous geographic and temporal diversity in water supply and demand situations suggest that no single institutional arrangement is likely to be preferred in all instances (Frederick, 1992). While Howe et al. (1986) have argued that markets meet all the criteria for effective water allocation better than any likely alternative, this is true for Indian situations (Kumar, 2003).

The absence of well-defined property rights regimes is a major source of uncertainty about the negative environmental impacts of resource use, leading to inefficient and sustainable use (Kay et al., 1997). This has been evident in the case of both groundwater and canal water supplied for irrigation. In the Indian context, many researchers in the recent past have suggested the establishment of property rights as a means of building the institutional capability to ensure equity in allocation and efficiency in use of water across sectors (Kumar, 2000a). But, if the rights are allocated only to the use of water, it can create incentives to use it even when there is no good use of it (Frederick, 1992). Therefore, water rights have to be tradable (IRMA/UNICEF, 2001; Kumar and Singh, 2001). Establishing privately owned property rights that are tradable is critical to establishing conditions under which individuals will have opportunities and incentives to develop and use the resource efficiently, or transfer it to more efficient uses (Frederick, 1992).

Empirical evidence collected on the functioning of groundwater irrigation institutions in north Gujarat show that, under a system of fixed volumetric water use rights, farmers prefer to grow highly water-efficient crops (Kumar, 2005, 2010). Tradable private property rights need to be enforced for groundwater and water supplied from public reservoirs for irrigation. In the case of groundwater and canal water supplied for irrigation, as individuals enjoy access to the resource, it is envisaged that private property rights for individual users will be created.

Fixing norms for the allocation of volumetric water rights across individual sectors, namely agriculture, industry and the domestic front, should involve considerations such as the physical sustainability of the water resource system and environmental sustainability. The total water allocated from any region/basin, therefore, should not exceed the difference between the annual renewable freshwater and the ecological demand, or the utilizable freshwater – whichever is less. In accordance with such norms, in those regions in which water resources are abundant by nature, such as the eastern part of UP, Bihar, Odisha and West Bengal, the volumetric water rights of individual sectors and users, especially farmers, would be very high. In these regions, land availability would continue to be an important factor in deciding returns from agriculture (Kumar, 2003). The farmers will, therefore, have to choose crops which are more water intensive and which would encourage intensive use of the same piece of land. In states like Bihar and UP, water rights would not mean much for a large number of cultivators, who have marginal holdings or no land.

In such situations, the allocation norms in agriculture need to be carefully designed, if equal opportunities to improve their own farm economies are to be given to all types of cultivators. In water allocation, the food security needs of the families could be given priority, rather than the farm size. This will result in a disproportionate allocation of rights in favour of small and marginal farmers. This can induce interlocked land, pump and water markets, wherein the rich well-owning farmers will offer pump services to farmers who do not have their own irrigation sources, and can, in return, use a portion of their water rights. This will force the rich well owners to charge less for the pump irrigation services they provide, thereby promoting greater equity in access to groundwater in these regions. They may also enter into sharecropping arrangement with the landless. As they are likely to have excess water, a good economic opportunity lies for landless, small and marginal farmers in transferring water in bulk to water-scarce regions, or those cities and industrial areas which are concentrated points of large demand for water. The physical conditions for transfer of water from rich areas to water-scarce areas exist in many regions (Kumar and Singh, 2001; Kumar, 2003).

Encouraging the efficient use of water in agriculture

Pricing of irrigation water

In spite of the recommendations of the Second Irrigation Commission, state irrigation bureaucracies have failed to raise those water charges which make economic sense, due to the potential social and political ramifications. This failure has its roots in the absence of institutional capability to improve the quality of irrigation services and correctly monitor water use; and the lack of institutional arrangements at the lowest level to recover water charges from individual farmers and enforce penalties on free-riders. A few successes have been seen in areas where the PIM programme is implemented, in that farmers there have shown willingness to pay more to the Water User Associations for irrigation services.

In the recent past, there has been significant debate over the usefulness of irrigation water pricing as a way to regulate water demand, with some arguing for (Tsur and Dinar, 1995), and some others arguing against by pointing out shortcoming at both theoretical and practical levels (Perry, 2001). There are three major and important contentions of those who argue against pricing: a) they question the logic in the proposition that "if the marginal costs are nil, farmers would be encouraged to use large quantities of water before its marginal productivity becomes zero, consuming much more than the accepted standards and needs" (Molle and Turral, 2004); b) the demand for irrigation water is inelastic to low prices, and the tariff levels at which the demand becomes elastic to price changes would be so high that it becomes socially and politically unviable to introduce (Perry, 2001) and c) there are no reasons for farmers to use too much water, which can cause over-irrigation (Molle and Turral, 2004).

However, as noted by Kumar (2010), these arguments have weaknesses. The most important issue is in linking irrigation charges and demand for water (see Perry, 2001). Merely raising the water tariff without improving the quality and reliability of irrigation will not only make little economic sense, but also would find few takers. As returns from irrigated crops are more elastic to the quality of irrigation than its price (Kumar and Singh, 2001), a poor quality of irrigation increases farmers' resistance to pay for the irrigation services they receive. Therefore, the water "diverted" by farmers in their fields does not reflect the actual demand for water in a true economic sense, so long as they do not pay for it. In other words, the impact of tariff changes on irrigation water demand can be analyzed only when the water use is monitored and farmers are made to pay for the water on volumetric basis. The above arguments also lead us to the conclusion that the rates for canal water can be increased to substantially higher levels, provided the quality of irrigation water is enhanced. But water pricing for irrigation can impact poor farmers adversely, if pitched at higher levels (Frederick, 1992). One of the ways of reducing the negative impact on access equity is to introduce a progressive pricing system. An appropriate pricing structure for water followed by clearly recognized private property rights and a good quality irrigation service could help achieve the desired effect of pricing changes on demand management. It also means that, if positive marginal prices are followed by improved quality, the actual (aggregate) demand for irrigation water might actually go up, depending on the availability of land and alternative crops that give a higher return per unit of land. This is because the tendency of the farmers would be to increase the volume of water used to maintain or raise their net income (Kumar and Singh, 2001). Hence, water rationing is important to demand regulation in most situations (Perry, 2001).

Pro rata pricing of electricity in the farm sector

In the past, many researchers have suggested rational pricing of electricity as a potential fiscal tool for sustainable groundwater use in India (Moench 1995; Saleth, 1997; Kumar, 2005; Kumar et al., 2011b). Many argue that a flat rate based pricing structure in the farm sector creates an incentive for farmers to over-extract it, as the marginal cost of extraction is zero. However, empirical evidence does not seem to suggest any impact of the cost of extraction on the use of groundwater for irrigation in water-scarce areas (Kumar and Patel, 1995), or in areas where water charges reflect the scarce value of the commodity (Kumar, 2005). The policies with regard to water and electricity pricing are guided by strong political and economic considerations (Moench, 1995). But the recent past has seen some remarkable success in introducing metering, and charging a power tariff based on actual consumption, in West Bengal (Mukherji et al., 2009) and Gujarat (Kumar et al., 2011b). In Punjab, the farmers have been crying foul over the deteriorating power supply, which is free, and instead were demanding a good quality power supply with a price. A field survey carried out in Haryana showed farmers' willingness to pay a higher tariff for electricity if the quality of supply is improved (Kumar et al., 2011b).

Studies carried out in the Mehsana district of north Gujarat and coastal Saurashtra on diesel and electric well commands show that control over watering will have a greater bearing on the net returns from irrigation than the cost of irrigation. This means that the desired impacts of changes in the pricing structure of electricity on the economic efficiency of irrigated crops can be realized only if the quality of power supply is ensured (IRMA/UNICEF, 2001). Kumar (2005) showed that unit pricing of electricity influences groundwater use efficiency and productivity positively. It also shows that the levels of pricing at which demand for electricity and groundwater becomes elastic to tariff are socio-economically viable. Furthermore, the water productivity impacts of pricing would be highest when water is volumetrically allocated with rationing. This evidence build a strong case for introducing pricing changes in the electricity supplied to the farm sector. One of the arguments against price change is that the higher marginal cost of supplying electricity under a metered system could reduce net social welfare, as a result of the reduction in a) demand for electricity and groundwater; and b) the net surpluses individual farmers could generate from cropping. Another argument against using pricing is that, for power tariff levels to be in the responsive region of the power demand curve, prices are often so high that it may become socially unviable. The analyses to which we have referred questioned the validity of these arguments.

Kumar (2005) showed that higher demand reduction in groundwater and electricity would be achieved if volumetric rationing of energy/water was done in tandem with an induced marginal cost of using energy/water. Furthermore, the effectiveness in implementing pricing policies would depend heavily on the ability to supply high quality in rural areas and meter its use, levy the charges without default, prevent thefts, and penalize free-riders. But, in areas with abundant and shallow groundwater, especially in eastern India, the electricity pricing structure should be such that it encourages greater exploitation of groundwater (GOI, 2008).

Nevertheless, even full cost pricing of electricity might work out to be cheaper for farmers, as the depth of pumping is low in these regions.

Kumar et al. (2011b) – based on an analysis of data collected from electric well owners, buyers of water from electric wells, diesel well owners, and buyers of water from diesel wells in South Bihar and eastern Uttar Pradesh; and electric well owners under both pro rata tariff and flat rate tariff in north Gujarat – showed that a pro rata power tariff and increase in unit charges for electricity would encourage farmers to use water more efficiently at the farm level.[4] They do this by modifying their farming systems through the selection of high valued and water-efficient crops and livestock, and the efficient use of irrigation water and other inputs for individual crops in their farm. This also indicates that their willingness to take risks will be greater under such pricing regimes, as most high valued crops are also risky crops.

The study showed that the biomass output and net income per unit volume of water was higher for water buyers and farmers who had metered connections. Also, the net return for crops per unit area of land was higher. The overall outcome was that their farm level water productivity (INR per cubic metre) was higher as compared to well owners who paid for electricity on the basis of connected load. They were also found to use less groundwater per unit of irrigated land. Furthermore, the farmers who bought water from diesel well owners, and who incurred the highest cost for irrigation water, obtained a higher return per unit volume of water when compared to diesel well owners and buyers of water from electric well owners. These findings together indicated that pro rata pricing of electricity with high energy tariff would be socio-economically viable.

Thus, the empirical studies carried out so far on the issue of energy pricing on groundwater use in India show that the introduction of consumption-based pricing of electricity and an increase in unit charges, if combined with improvement in the quality of power supply, will lead to greater agricultural income and a reduction in use of groundwater. Contrary to what has been widely believed, it will not result in any adverse impact on the economic prospects of irrigated farming, and net income from farming is less elastic to the price of irrigation water than its quality and reliability. With the advent of pre-paid electronic meters which work through scratch cards (Zekri, 2008) and work on internet or mobile technology and remotely-sensible meters (Mukherji et al., 2009), the transaction cost of metering can be minimized to a great extent.

The pre-paid meters are ideal to enable remote areas to monitor energy use and control groundwater use online from a centralized station (Zekri, 2008). Over the past 7–8 years, there has been a remarkable improvement in the quality of services provided by internet and mobile (satellite) phone services, especially in rural areas, with a phenomenal increase in the number of consumers (Kumar et al., 2011b). Such technologies are particularly important when there are large numbers of agro wells, and the transaction cost of visiting wells and taking meter reading is likely to be very high (Zekri, 2008). It will be inevitable in rural India. Pre-paid meters prevent electricity pilferage through manipulation of pump capacity. They can be operated through tokens, scratch cards, magnetic cards or they can be recharged digitally through the internet and SMS.

The use of remotely-sensible meterscan also avoid the huge transaction cost of metering. The technology used in these meters enables them to be installed in places where tampering by farmers and meter readers will be difficult, yet where readings can be easily obtained. This is now used for measuring electricity consumption by agro wells in West Bengal (Mukherji et al., 2009).

But, such fiscal instruments are required in regions which experience over-draft. As we have seen in Chapter 2, in India, overdraft appears to be occurring in regions which experience low to medium rainfall with high aridity. Metering and pro rata pricing of electricity may not be required in those regions which have abundant groundwater, if the issue of cost recovery in electricity supply can be addressed through other modes of pricing. The reason is that metering is essentially an economic decision and the benefits of metering have to justify the efforts involved. In water-abundant regions, the social and economic benefits of groundwater conservation through metering may not be very significant. Therefore, in those regions, the pricing structure should be designed in such a way that it encourages greater use of groundwater for boosting agricultural production. But caution should be exercised, to see that it does not create negative effects on equity in distribution of energy subsidy benefits. The slab system of pricing, based on the connected load of the motor, can be a good basis for pricing electricity in such areas.

Future areas of action in agriculture

As we have argued in the previous section, strategies and policies for agricultural growth in India have to be region-specific, considering the unique problems each region poses and the opportunities each provides. The promotion of low cost, energy efficient water harnessing technologies, such as treadle pumps and micro diesel pumps, through the supply of information, materials and services, can enable poor farmers in the agriculturally-backward eastern and north-eastern parts of our country to access irrigation water. This will create millions of microeconomies, with sustainable utilization of water resources in the water-abundant regions (Postel, 1999; Shah et al., 2000). Low cost water saving technologies will enable the poorest sections of the communities to practise irrigated agriculture with very limited water in water-scarce regions. Land-based interventions for drought-proofing, water harvesting and artificial recharge under NREGA should be planned carefully, after proper consideration of hydrological and economic aspects, so as to improve the overall water economy and to reduce the negative welfare effects on society (Bassi and Kumar, 2010).

In naturally water-scarce regions, where such interventions are likely to create negative welfare effects, funds should be earmarked for providing subsidies for micro-irrigation (Bassi and Kumar, 2010). In those regions, welfare gains would be substantial, for the following reasons: a) real water saving through micro-irrigation would be possible; and b) the value of saved water would be high in these regions, owing to its physical scarcity. This would, to an extent, help reduce groundwater over-draft in some of these regions, along with raising crop outputs, whereas in other regions, the impact would be only increased agricultural output.

If India has to expand its irrigation potential for agricultural growth on a sustainable basis, and sustain intensive well irrigation in the naturally water-scarce regions, the transfer of water from water-abundant regions to water-scarce regions would be essential. Inter-basin or inter-regional water transfer is a natural choice for irrigation expansion and groundwater replenishment. The water-scarce regions have the natural advantage of being able to produce more crops, if irrigation water is available, as a sufficient amount of arable land remains uncultivated and unirrigated. In contrast, water-abundant regions face an acute shortage of arable land, in addition to the ecological constraints to crop production induced by floods. This calls for the design of an entirely new water management system which would promote efficient use of water in agriculture and the improved recharge of groundwater at no extra cost.

In the alluvial areas of western and north-western India, the depleted aquifers can be effectively replenished through diversion of water for irrigation during rainy season, with a large amount of land available for infiltration and good hydraulic conductivity in the de-watered zone. In contrast, in the hard rock areas of peninsular India, the hundreds of thousands of tanks which dot the region could be integrated with the infrastructure which transfers the water to this region. As many of them do not receive sufficient inflows from their catchments, they can store a portion of the imported water, while the remaining water could be diverted to agricultural land.

Enforcement of private and tradable water rights in groundwater and water supplied from public reservoirs can together bring about a significant increase in farm outputs, with a reduction in aggregate demand for water in agriculture. It will also bring about more equitable access to, and control over, the water available from canals and groundwater for food production and ensure household-level food security. This has to be complemented by the volumetric pricing of canal water, and pro rata pricing of electricity in the farm sector, with improved quality and reliability of the supplied power. Metering and pro rata pricing of electricity has to receive priority in naturally water-scarce regions which also experience groundwater over-draft. In groundwater-abundant regions, however, the pricing structure should be designed in such a way that it encourages greater use of groundwater for agricultural production.

Notes

1. This is due to extremely low rural electrification and the high cost of fuel to run diesel engines.
2. The only exception is the Godavari river basin, passing through Maharashtra and Andhra Pradesh, which is water-abundant.
3. Thisis a process by which surplus runoff, or surface water during times of surplus, is stored in aquifers for use in times of shortage. Water can also be over-drawn, onthe understanding that it willbe replenished later (Hosteler, undated).
4. Here, buyers of irrigation water, who pay for irrigation services on an hourly basis, are considered to be the proxy for the pro rata pricing of electricity. This is because the effect of both pro rata pricing of electricity and volumetric irrigation water charges, in terms of marginal cost of irrigation, would be the same.

References

Amarasinghe, U., Shah, T., & Singh, O. P. (2007). *Changing consumptions patterns: Implications on food and water demand in India.* Research Report 119. Colombo, Sri Lanka: International Water Management Institute.

Ambasta, P., Vijay Shankar, P. S., & Shah, M. (2008). Two years of NREGA: The road ahead. *Economic and Political Weekly, 43* (8), 41–50.

Bassi, N., & Kumar, M. D. (2010). *NREGA and rural water management: Improving the welfare effects.* Occasional Paper 3. Hyderabad, India: Institute for Resource Analysis and Policy.

Chowdhury, A. (2011). Food price hikes: How much is due to excessive speculation? *Economic and Political Weekly, 46* (28), 12–15.

Contor, B. (2009). *Groundwater banking and the conjunctive management of groundwater and surface-water in the Upper Snake River basin of Idaho.* Idaho, USA: Idaho Water Resources Research Institute.

Frederick, K. D. (1992). *Balancing water demand with supplies: The role of management in a world of increasing scarcity.* Technical Paper 189. Washington, DC, USA: The World Bank.

Government of India. (2008). *11th Five Year Plan 2007–2012. Vol. III: Agriculture, Rural Development, Industry, Services, and Physical Infrastructure.* New Delhi, India: Planning Commission, Government of India.

Hostetler, S. (undated). *The Australian Water Bank: The bank we have to have?* Canberra, Australia: Integrated Water Sciences, Bureau of Rural Sciences, The Australian Government.

Howe, C.W., Schurmeier, D. R., & Shaw, D. W. Jr. (1986). Innovative approaches to water allocation: The potential for water markets. *Water Resources Research, 22* (4), 439–445.

Institute of Rural Management Anand & UNICEF. (2001). *White paper on water in Gujarat.* Report submitted to the Government of Gujarat. Anand, Gujarat, India: Institute of Rural Management.

Kay, M., Franks, T., & Smith, L. (1997). *Water: Economics, management and demand.* London, UK: E. & F. N. Spon.

Kijne, J., Randolph, B., & Molden, D. (2003). Improving water productivity in agriculture: Editors' overview. In J. Kijne, R. Barker & D Molden (Eds.), *Water productivity in agriculture: Limits and opportunities for improvement.* Comprehensive assessment of water management in agriculture series1. Wallingford, UK: CABI Publishing in association with the International Water Management Institute.

Kumar, M. D. (2000a). Institutional framework for managing groundwater: A case study of community organisations in Gujarat, India. *Water Policy, 2* (6).

Kumar, M. D. (2000b). *Irrigation with a manual pump: Impact of treadle pump on farming enterprise and food security in coastal Orissa.* Working paper 148. Anand, Gujarat, India: Institute of Rural Management Anand.

Kumar, M. D. (2001). *Demand management in the face of growing scarcity and conflicts over water use in India: Future options.* Working paper 153. Anand, Gujarat, India: Institute of Rural Management Anand.

Kumar, M. D. (2003). *Food security and sustainable agriculture: India's water management challenge.* Working Paper 60. Colombo, Sri Lanka: International Water Management Institute.

Kumar, M. D. (2005). Impact of electricity prices and volumetric water allocation on groundwater demand management. *Energy Policy, 33* (1).

Kumar, M. D. (2007). *Groundwater management in India: Physical, institutional and policy alternatives.* New Delhi, India: Sage Publications.

Kumar, M. D. (2009). *Water management in India: What works, what doesn't.* New Delhi, India: Gyan Books.

Kumar, M. D. (2010). *Managing water in river basins: Hydrology, economics, and institutions*. New Delhi, India: Oxford University Press.

Kumar, M. D., & Patel, P. J. (1995). Depleting buffer and farmers' response: Study of villages in Kheralu, Mehsana, Gujarat. In M. Moench (Ed.), *Electricity prices: A tool for groundwater management in India?* Monograph. Ahmedabad, India: VIKSAT & Natural Heritage Institute.

Kumar, M. D., & Singh, O. P. (2001). Market instruments for demand management in the face of growing scarcity and overuse of water in India. *Water Policy*, *5* (3), 86–102.

Kumar, M. D., & Singh, O. P. (2005). Virtual water in global food and water policy making: Is there a need for rethinking? *Water Resources Management*, *19*, 759–789

Kumar, M. D., Patel, A., Ravindranath, R., & Singh, O. P. (2008a). Chasing a mirage: water harvesting and artificial recharge in naturally water-scarce regions. *Economic and Political Weekly*, *43* (35), 61–71.

Kumar, M. D., Turral, H., Sharma, B., Amarasinghe, U., & Singh, O.P. (2008b). Water saving and yield enhancing micro irrigation technologies: When do they become best bet technologies? In M. D. Kumar (Ed.), *Managing water in the face of growing scarcity, inequity and declining returns: Exploring fresh approaches*. Proceedings of the 7th annual partners' meet, IWMI-Tata Water Policy Research Program, ICRISAT, Patancheru, 2–4 April, 2008.

Kumar, M. D., Kumar Malla, A., & Kumar Tripathy, S. (2008c). Economic value of water in agriculture: comparative analysis of a water-scarce and a water-rich region. *Water International*, *33* (2), 214–230.

Kumar, M. D., & Bassi, N. (2011). Maximizing the social and economic returns from Sardar Sarovar Project: Thinking beyond convention. In R. Parthasarathy, & R. Dolakhya (Eds.), *Sardar Sarovar Project on the River Narmada: Impacts so far and ways forward* (pp. 747–776). Delhi, India: Concept Publishing.

Kumar, M. D., Scott, C., & Singh, O. P. (2011b). Inducing the shift from flat rate or free agricultural power to metered supply: Implications for groundwater depletion and power sector viability. *Journal of Hydrology*, *409* (2011), 382–394.

Moench, M. (Ed.). (1995). *Electricity pricing: A tool for groundwater management in India?* Monograph. Ahmedabad, India: VIKSAT & Natural Heritage Institute.

Molle, F., & Turral, H. (2004). *Demand management in a basin perspective: Is the potential for water saving overrated?* Paper presented at the international water demand management conference, June 2004, Dead Sea, Jordan.

Mukherji, A., Das, B., Mujumdar, N., Nayak, N. C., Sethi, R. R., Sharma, B. R., & Banerjee, P. S. (2009). Metering of agricultural power supply in West Bengal: Who gains and who loses? *Energy Policy*, *37* (12), 5530–5539.

Nandakumar, T., Ganguly, K., Sharma, P., & Gulati, A. (2010). *Food and nutrition security status in India: Opportunities for investment partnerships*. ADB sustainable development working paper series 16. Manila, Philippines: Asian Development Bank.

National Council of Applied Economic Research & Public Interest Foundation. (2009). *NCAER-PIF study on evaluating performance of National Rural Employment Guarantee Act*. New Delhi, India: National Council of Applied Economic Research.

Panagariya, A. (2009). More bang for the buck. *The Times of India*, 19 September 2009.

Perry, C.J. (2001). Water at any price? Issues and options in charging for irrigation water. *Irrigation and Drainage*, *50*, 1–7.

Postel, S. (1992). *The last oasis: Facing water scarcity*. World Watch environmental alert series. London, UK: Earthscan.

Postel, S. (1999). Pillars of sand: Can the irrigation miracle last? Washington, DC, USA: W.W. Norton & Co. Ltd.

Postel, S., Polak, P., Gonzales, F., & Keller, J. (2001). Drip irrigation for small farmers: A new initiative to alleviate hunger and poverty. *Water International, 26* (1),3–13.

Saleth, R. M. (1997). Power tariff policy for groundwater regulation: Efficiency, equity and sustainability. *Artha Vijnana, 39* (3).

Saleth, R. M., & Dinar, A. (1999). *Water challenge and institutional responses (across country perspective)*. Policy research working paper series 2045. Washington, DC, USA: The World Bank.

Shah, T., Alam, M., Kumar, M. D., Nagar, R. K., & Singh, M. (2000). *Pedaling out of poverty: Social impact of a manual irrigation technology in South Asia*. Research Report 45. Colombo, Sri Lanka: International Water Management Institute.

Tiwari, R., Somashekhar, H. I., Parama, V. R. R., Murthy, I. K., Mohan Kumar, M. S., Kumar, B. K. M., Parate, H., Varma, M., Malaviya, S., Rao, S. A., Sengupta, A., Kattumuri, R., & Ravindranath, N. H. (2011). MGNREGA for environmental service enhancement and vulnerability reduction: Rapid appraisal in Chitradurga district, Karnataka. *Economic and Political Weekly, 46* (20), 39–47.

Tsur, Y., &Dinar, A. (1995). *Efficiency and equity considerations in pricing and allocating irrigation water*. Policy research working paper series 1460, Washington, DC, USA: The World Bank.

Watt, J. (2008). *The effect of irrigation on surface-groundwater interactions: Quantifying time dependent spatial dynamics in irrigation systems*. Doctorial dissertation, Charles Sturt University School of Environmental Sciences.

Zekri, S. (2008). Using economic incentives and regulations to reduce seawater intrusion in the Batinah coastal area of Oman. *Agricultural Water Management, 95* (3).

10 Investment strategies and technology options for sustainable agricultural development in Asia

Challenges in the emerging context

P. K. Viswanathan, M. Dinesh Kumar and M. V. K. Sivamohan

Introduction

The changes in agriculture trade policies in the post-WTO context proved detrimental to most developing economies, especially in the Asian region. One reason is that a large number of these countries started re-orienting their agriculture development policies towards growing non-food or commercial/high value crops. The other reason is that they opt for non-agricultural land use systems that are driven by market forces. In fact, this shift in agricultural policies has resulted in serious food security problems for most of these countries. The severity of the food security challenge became more evident after 2008, when the global economy was severely affected by the financial crisis. The magnitude of the problem was intensified by the crisis, with several developing countries experiencing a severe shortage of food and unprecedented escalation in food prices. During the first five months of 2008, food prices rose from US $385 per tonne during January 2008 to US $873 (US $1 equals INR 50) per ton in May 2008 (World Bank, 2008).

Although the rise in food prices was a transient phenomenon, its consequences exacerbated concerns about the severity of global food insecurity. While many countries have been facing a steep decline in the rate of growth in food production, the demand for food has been on the increase, due to growing human and livestock populations. These concerns are further intensified by the possibilities of agriculture and food production systems being severely affected by likely changes in climate. The declining growth in global food production, the financial and economic slowdown, and climatic vulnerabilities are crucial factors. Furthermore, land acquisition by industries and infrastructure projects as seen in large parts of the developing world, seem to have undermined the food security of 75–105 million new poor globally (World Bank, 2008; Woolverton et al., 2010).

Ironically, countries in the Asian region, which are also the major suppliers of food to the world (especially to the industrially advanced countries), have been seriously affected by the nascent food insecurity challenges. Estimates by the Food and Agriculture Organization (FAO) indicate that between 2003 and 2005, 541.9

million people in the region were undernourished (FAO, 2009), and, by the end of 2010, the Asian countries will account for about one-half of the world's undernourished population, of which two-thirds will be from South Asia (Mittal and Deepthi, 2009). Further evidence reveals that, despite public policies and intervention to enhance food production and distribution in the majority of these countries, the overall improvement in nutritional status has been very slow (FAO, 2008a). Moreover, there exists chronic under-nourishment in the low income groups, constituting about half of the population, and is particularly affecting the children, women and the old. It has also been observed by many that the proportion of consumption expenditure on food has been steadily going down even among the households with chronic under-nourishment. Thus, despite achieving commendable levels of food production, many Asian countries have failed to address the problem of mass under-nourishment in an effective way (Meinzen-Dick et al., 2011). These eventualities suggest that household food security is more about the income security that provides access to food, rather than just the availability of food. In other words, it is the inability of the households to purchase food or have access to the physical storage of food that leads to the household's food insecurity (FAO, 2008a).

Incidentally, liberalization also raises several challenges as regards the significance and effectiveness of macro policy interventions and the safety net programmes enunciated by developing countries, which are aimed at enhancing food availability. Building buffer stocks is one such policy intervention. On the other hand, there are strong indications of dwindling public investment in agriculture, particularly in the development of irrigation infrastructure across all countries in the Asian region. These challenges signal the severity of the agrarian crisis looming large in the regional context of Asia, with long-term implications for food security, sustained investment in revamping the irrigation systems and management of farm lands. These challenges are further magnified by the growing environmental problems caused by agricultural development, namely a) degradation of farm lands resulting from intensive use of chemical inputs; and b) depletion of water resources and their quality deterioration (Thapa et al., 2010).

This chapter explores the need for, and challenges involved in, revamping Asian smallholder agriculture in the wake of dwindling public investment in agriculture, on the one hand, and the challenges emerging from the increasing vulnerability of farmers to food insecurity (as a result of resource degradation problems and climatic variability) on the other. In particular, the chapter discusses the persistent challenges facing sustainable farm production systems in Asia, as emerging from the structural transformation experienced by the majority of countries. This apart, the competing demands of growing urbanization/industrialization and ecosystem conservation pose a newer set of challenges. Devising new institutional models of investment for the development and modernization of agriculture and irrigation systems is necessary in order to strengthen and enhance the capabilities of the developing countries to sustain rural transformation.

The chapter is organized into four sections. The second section provides a brief comparative overview of agricultural development in major Asian countries, considering, especially, the impact of Green Revolution technologies, public policies and investments in achieving self-sufficiency in food production. The third section examines some critical aspects of the agrarian transformation and their implications for food security, as well as considering the challenges facing the development and effective utilization of land and water resources in the region. The fourth section explores the need to strengthen the capacities of the agriculture sector in Asia – in terms of revamping the institutional, as well as technological, domains that facilitate investment in agriculture and water resources development – in order to address current and future food security challenges. The chapter uses a comparative analytical perspective with respect to five major Asian countries – India, Bangladesh, Vietnam, South Korea (Korea, henceforth) and Thailand – which are distinct in terms of the emerging challenges confronting investment in agriculture and water sector development.

Asian agriculture development under the Green Revolution: the role of policies, technologies and institutions

There is no denying of the fact that Asian agriculture underwent tremendous transformation between the 1960s and 1980s, owing to the policies and institutional development strategies that evolved under the Green Revolution (GR) regime. During the 1960s, when a vast segment of the Asian population faced famine and starvation, the GR technologies came as a boon to many countries, and, as a result, many have vigorously adopted these technologies (HYV seeds, fertilizers, pesticides, mechanization, etc.). The injection of massive public investment for development of water resources, along with input subsidies (seeds, fertilizers, energy, irrigation) and credit support, have all enabled the region's economies to overcome the food deficit, with many transforming into surplus producers in less than a decade (Viswanathan et al., 2012). Farmers, particularly small and marginal farmers, had fast adopted these technologies, which were developed at international agricultural research centres, such as the International Rice Research Institute (IRRI) in the Philippines, the International Maize and Wheat Research Centre (CIMMYT) in Mexico and the World Vegetable Centre in Taiwan (AVRDC) (Kaosa-ard, Santikarn and Rerkasem, 2000; Chand, 2010). Furthermore, between 1961 and 2002, the irrigated area in Asia more than doubled, as governments sought to achieve food self-sufficiency, improve welfare and generate economic growth. Investment in irrigation provided the key to unlocking Asia's agricultural potential at that time. For example, in South Asia cereal production rose by 137 per cent from 1970 to 2007, and this was achieved using only 3 per cent more land (Mukherji et al., 2009).

It is apparent that the adoption of GR technologies significantly benefited all five countries, in terms of increased production of food grains, especially during the first three decades. For instance, between 1961 and 1975, production of food grains increased by more than 50 per cent in India and Thailand, by about 50 per cent in Vietnam, and by 41 per cent in Bangladesh. Korea was an exception with a decline in food production after the II period (see Table 10.1).

Table 10.1 Trends in the production of food grains, 1961 to 2009

Country	*Average annual production (million tonnes)*			*Per cent change between*	
	I period (1961–1975)	*II period (1976–1990)*	*III Period (1991–2009)*	*I & III period*	*II & III period*
Bangladesh	16.27	22.80	36.17	122.3	58.6
India	100.85	156.57	228.46	126.5	45.9
Korea	7.59	8.98	7.21	−5.0	−19.7
Vietnam	10.02	14.83	33.02	229.5	122.7
Thailand	14.31	21.74	29.91	109.0	37.6

Source: Viswanathan et al., 2012.

However, it may be noted that much of the contribution to growth in food production came from a single crop, i.e., rice, which also has been the largest benefactor of GR in all of the countries except India. The contribution of rice to the total food grain production is as high as 96 per cent in Korea, 95 per cent in Bangladesh, 89 per cent in Vietnam and 88 per cent in Thailand. In contrast, India has strategically diversified its food production by growing rice and wheat and, hence, contribution from rice in total food production has been 56 per cent, followed by wheat with 30 per cent; this has taken advantage of the variations in agro-climate across regions.

The yield impact of GR technologies has been quite significant in the case of rice, though with notable variations across regions. For instance, Korea achieved the highest yield levels, in the range of 4.0–4.8 tonnes per ha over the last five decades (1961–2009). In Vietnam, rice yield varied from 1.92 to 4.8 tonnes per ha and in Bangladesh it varied from 1.68 to 3.71 tonnes per ha during the same period. India and Thailand had relatively lower levels of rice productivity, in the range of 1.48 to 3.1 tonnes per ha and 1.78 to 2.82 tonnes per ha, respectively.

Besides benefiting from GR technology, the countries also made significant investments in research and development, infrastructure development and in extension programmes for increasing the production of commercial crops, mainly for export. Vietnam and India were already growing tropical cash crops, such as rubber, tea and coffee, even during the pre-war/colonial period, so as to boost export earnings. Thus, the GR period also coincided with the dynamic growth of commercial agriculture in these countries. Nevertheless, despite conscientious efforts, the promotion of cash crops did not result in a large-scale diversification of agriculture, especially in Bangladesh and Vietnam, as evident from the continued dominance of food crops (mainly rice) in the gross cropped area in these countries (86 per cent and 72 per cent, respectively). In contrast, India, Korea and Thailand were more successful in crop diversification, as the area under non-food/commercial crops was about 44–47 per cent of the gross cropped area during 2007.

Most of the growth in these countries can be attributed to increases in yield per unit of land facilitated by investments in technology. The increase in crop yields was a cumulative outcome of a wider adoption of new plant varieties, developed primarily by public plant-breeding institutes, and the increased use of fertilizer and irrigation. In the livestock sector, growth in animal productivity was attributable to the combination of new breeds of poultry and piggery, developed primarily by the private sector, with new feed, health and commercial management practices also being developed by private firms.

It may be argued that the growth and spread of GR technologies in Asia has been mainly triggered by the public sector investments in agriculture development, research and development and extension. On the other hand, the growth of commercial farming got stimulus from investment by the private sector in research on enhancing the yields of corn, sunflower, pearl millet, sorghum and cotton in India; corn and horticultural crop yields in Thailand; corn in the Philippines; and corn and tobacco in Pakistan. In fact, in both the cases, the complementary nature of the research and development work between the public and private sectors has been quite apparent. In the case of India, the establishment of public research institutes in the country and the international centres provided a considerable boost to private plant-breeding research (Pray et al., 1998).

The growth of food crops, along with the expansion of commercial farming, has enabled these five countries to gain significantly from increased export earnings over time. As evident from Figure 10.1, Thailand had the highest gains in agricultural exports from US $13,126 million to US $20,716 million between 2004–06 and 2007–08. In relative terms, India's agricultural exports increased by 82 per cent, followed by Vietnam (69 per cent), Thailand (58 per cent), Bangladesh (40 per cent) and Korea (28 per cent) during the six year period 2004–09. However, all countries (except Vietnam) experienced a drop in exports between 1995–97

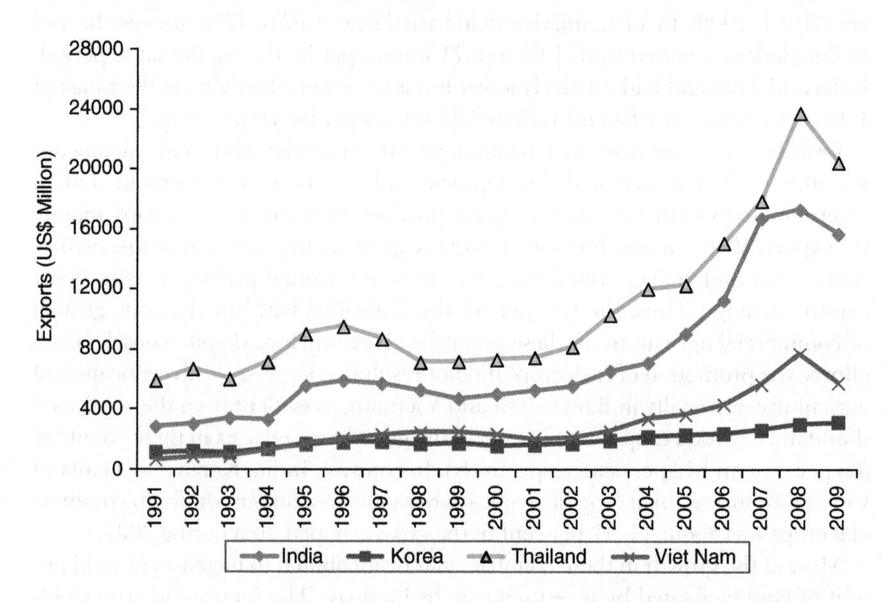

Figure 10.1 Trends in the agricultural exports of major Asian countries.

Source: Viswanathan et al., 2012, p. 44.

and 1998–00, due to the economic slowdown caused by the 1997 Asian financial crisis.

Nevertheless, there is no conclusive evidence to support the idea that small farmers in these countries have benefited from the upsurge in exports, as trends in the trade of food and food products convey a different story. All the countries have experienced significant decline in the export of food and food products, with an unfavourable balance of trade (BOT) in the food sector, especially in the case of Bangladesh and Korea. Thus, it becomes evident that the export growth observed was primarily driven by the rise in export of non-food/commercial products.

A closer look at the sectoral composition of GDP also indicates a significant structural transformation across the five countries, characterized by a drastic decline in the share of agriculture towards national income but with an increase in the share of the industry and services sectors (Table 10.2).

As evident from Table 10.2, there was significant increase in agricultural GDP in all of the countries, with India showing the highest increase in absolute terms (from US $80,351 million during 1985–89 to US $146,735 million during 2000–08). The decline in the share of the agriculture sector in the GDP has been more pronounced in case of Korea (74 per cent), followed by Vietnam (47.6 per cent), Thailand (40 per cent), India (39 per cent) and Bangladesh (33.5 per cent). While Vietnam and Thailand have shown notable increases in the growth of the manufacturing sector, Korea, India and Bangladesh have witnessed the emergence of the services sector as the single largest source of growth over time. This trend suggests that structural transformation in these countries has seen a shift in growth from the agricultural to the services sector, without a significant growth in the manufacturing sector, including agro-industries.

Table 10.2 Trends in agricultural GDP and its sectoral composition across major countries, 1985 to 2010

Period	*Bangladesh*	*India*	*Korea*	*Thailand*	*Vietnam*
1. Agricultural GDP (current US$ millions)					
1985–89	7531	80351	16899	8344	8994
2000–08	12213	146735	25844	18184	11029
Per cent change	62.2	82.6	52.9	117.9	22.6
2. Agriculture value added (per cent)					
1980–89	31.6	32.0	13.4	17.4	41.6
2000–10	21.0	19.5	3.5	10.5	21.8
Per cent change	–33.5	–39.1	–73.9	–39.7	–47.6
3. Manufacturing value added (per cent)					
1980–89	21.2	26	39.5	32.5	25.9
2000–10	27.2	27.7	37.1	43.5	39.7
Per cent change	28.3	6.5	–6.1	33.8	53.3
4. Services value added (per cent)					
1980–89	47.2	42.0	47.1	50.1	32.5
2000–10	51.8	52.8	59.4	46.0	38.5
Per cent change	9.7	25.7	26.1	–8.2	18.5

Source: World Bank World Development Indicators, 2010 (compiled).

Agrarian transition and its implications on food security and sustainable livelihoods in Asia

The agrarian transformation, as discussed in the preceding section, highlights the paradox of reducing the importance of agriculture as a source of growth while farming remains the mainstay of rural livelihoods in Asia. For instance, the proportion of the population that is rural remains as high as 72 per cent in Bangladesh and Vietnam, 70 per cent in India and 66 per cent in Thailand (FAO, 2010). Furthermore, there is heavy dependence by these countries' rural populations on agriculture for livelihood: it is as high as 63 per cent in Vietnam, 49 per cent in India, 46 per cent in Bangladesh and 41 per cent in Thailand. This indicates that the dramatic growth of the manufacturing and service sectors has not been effective in creating adequate employment opportunities for the large rural population. However, Korea is an exception, as the farming population fell sharply from 45 per cent (14.4 million) in 1971 to just 7.4 per cent (3.5 million) of the total population in 2003 (Song, 2006, as also cited in FAORAP, 2006; Lee and Kim, 2010).

Fragmentation and incidence of landlessness

The incidence of continued dependence by the rural population on farming has serious implications for that population's food security, as well as for that of the Asian region as a whole. This is because the increased pressure on farm lands resulted in the fragmentation of holdings, along with the shrinkage of arable land which is caused by degradation and conversion of farm lands for urban and industrial uses. Currently, small and marginal holdings constitute as much as 87 per cent of the holdings in Bangladesh, followed by Korea (85.2 per cent), Thailand (84.2 per cent) and Vietnam and India (82 per cent each). The average farm size is far below one hectare in Vietnam (0.57 ha) and Bangladesh (0.73 ha) and slightly above 1.0 ha in Korea (1.37 ha) and India (1.33 ha) (FAO, 2009). Though Thailand is an exception, with an average farm size of 3.65 ha at the national level, more than 30 per cent of the households own farm lands below 1.6 ha (Phrek, 2010). The situation is much worse if arable land per capita is considered. As of 2007, the average per capita farm land was the lowest at 0.07 ha in Bangladesh, while the other four countries had a slightly higher farm size, ie., 0.42 ha in Thailand, followed by India (0.21 ha), Korea (0.18 ha) and Vietnam (0.15 ha) (Viswanathan et al., 2012).

Alongside the increasing fragmentation of holdings, all countries (except Korea) are currently facing a major problem – the high occurrence of landlessness in rural areas. The Agriculture Census of Bangladesh reveals that about 4.5 million households (15.63 per cent) were completely landless during 2008, of which 73 per cent (3.26 million) lived in rural areas (BBS, 2008). The incidence of landlessness is also acute in India, as revealed by the NSSO survey: more than 40 per cent of rural households do not own land and the inequality in land ownership worsened between the 48th (1992) and 59th (2003–04) NSS rounds (Rawal, 2008). Vietnam reported a fourfold increase in landless households from 1.15 per cent in 1994 to 4.05 per cent in 2006 (Chung and Dang, 2010).

Food security issues

A major outcome of land fragmentation, declining per capita arable land and growing landlessness has been the dwindling per capita food production, especially in Korea and India, as is evident from Table 10.3.

As regards India, per capita food production marginally increased between 1961 and 1990 (to be precise, by 0.86 per cent between 1961 and 1975 and by about 2 per cent between 1976 and 1990), but this was followed by a decline of 0.41 per cent between 1991 and 2008. While Bangladesh, Thailand and Vietnam have recorded notable increase in per capita food production, Korea's situation has become deplorable, as average food production had reached an all-time low at 156 kg per capita during 1991–08. The cases of Thailand and Vietnam are quite striking, as they have been able to significantly improve the per capita availability of food grains over time (Figure 10.2).

Table 10.3 Trends in per capita food production in major countries

Country	*Average annual per capita food production (kg) during*		
	1961–1975	*1976–1990*	*1991–2008*
Bangladesh	247 (–2.59[a])	232 (0.39)	252 (4.12)
India	192 (0.86)	212 (1.86)	226 (–0.41)
Korea	249 (–0.20)	228 (–4.89)	156 (–2.25)
Vietnam	244 (–2.47)	257 (5.99)	415 (11.48)
Thailand	405 (0.98)	432 (1.80)	468 (7.50)

Source: Viswanathan et al., 2012.

Notes

a. Parenthetic figures indicate linear trends growth rates during the period.

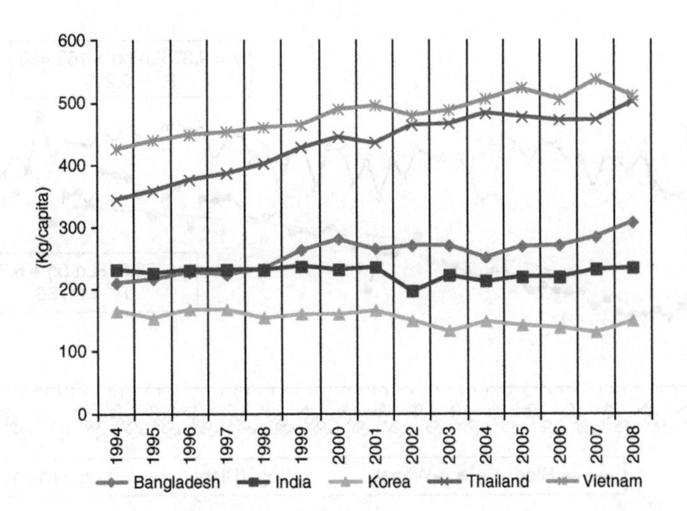

Figure 10.2 Trends in per capita food production in major countries, 1994–2008.

Source: World Development Indicators dataset, World Bank, 2008.

A closer look at the availability of food grains per capita in India reveals the declining trend for both rice and wheat since the beginning of the last decade (Figure 10.3). For instance, the average per capita availability of rice had declined by 7.8 per cent from 210 gram/day/capita during the 1990s to 193.8 gram/capita/day during the 2000s. However, the availability of wheat had only marginally (2 per cent) declined over the same period.

The availability of rice has also hown significant inter-annual fluctuations, as evident from Figure 10.3. As a matter of fact, the food security scenario in many regions in India is undermined by lack of access to food (entitlement failure, as argued by Amartya Sen) rather than under-production, as has emerged from some studies in the context of Gujarat (Chakravarty and Dand, 2006; Hirway and Mahadevia, 2003; Patnaik, 2009). Thus, what emerges from the analysis is that food insecurity in India, and in several other countries, including Thailand and Vietnam, in the Asian region, is a result of both entitlement failure and access failure.

Water security issues

An equally strong concern for many parts of Asia, along with the shrinkage of arable land, is the growing water scarcity. While water demand for irrigation has been growing, the increasing competition for the available water from domestic, municipal and manufacturing sectors and eco-system services is reducing the effective availability of water for crop and livestock production in many river basins. The growing water scarcity may have serious repercussions on future food production in most parts of the world (FAO, 2002; Mukherji et al., 2009). The current scenario of freshwater withdrawal in most parts of Asia is overwhelmingly skewed towards the irrigation sector.

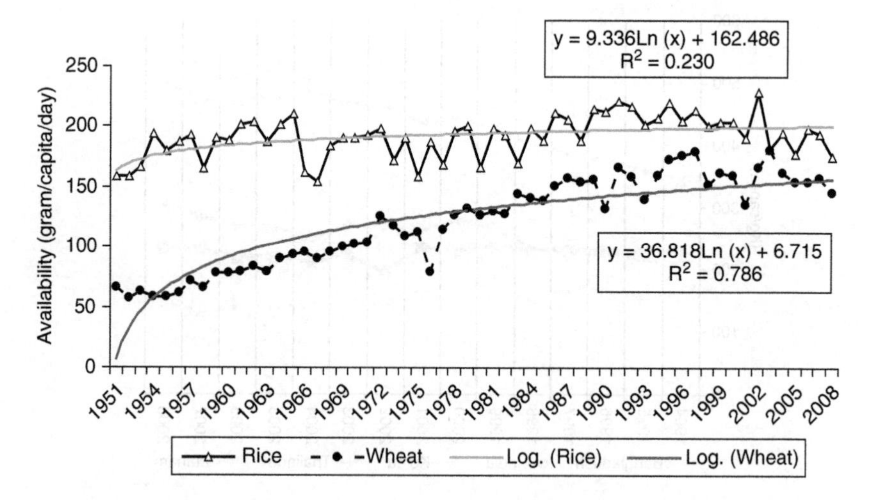

Figure 10.3 Long-term trends in the per capita net availability of food grains in India.

Source: World Development Indicators dataset, World Bank.

Table 10.4 Annual freshwater withdrawal for major uses, 2009

Country	*Agriculture (per cent)*	*Domestic (per cent)*	*Industry (per cent)*	*Freshwater use (billion m³)*	*Per cent share*	*Rice area irrigated*[a] *(per cent)*
Bangladesh	96.2	3.2	0.65	35.9	0.9	40.0
India	86.5	8.1	5.4	761.0	19.5	52.6
Korea	63.1	19.8	25.2	25.5	0.6	100.0
Thailand	95.1	2.5	2.5	57.3	1.5	25.0
Vietnam	68.1	7.7	24.1	82.0	2.1	53.0
South Asia	87.3	6.5	6.2	1026.6	26.3	58.6
World	75.2	9.5	15.3	3908.3	100.0	20.5

Source: World Development Report, 2011, World Bank (estimated).

Notes

a. Data relates to 2004–06.

As Table 10.4 shows, the share of the agriculture sector in freshwater withdrawal turns out to be as high as 96 per cent in Bangladesh and 95 per cent in Thailand, where the water use in the industrial and domestic/drinking water sectors has been almost negligible. Not only the percentage, but also the aggregate amount of freshwater used in agriculture is also very high in these regions, from a high of 761 BCM in India to 25.5 BCM in Korea. The case of India is quite distinct as it has the largest share (19.5 per cent) of freshwater withdrawal in the world (the share of South Asia being 26.3 per cent) along with the dominance of agriculture in freshwater withdrawal (86.5 per cent). It is also important to consider that the predominance of rice cultivation in the Asian region (58.6 per cent) exerts severe pressure on the scarce water resources. Notably, India is also not an exception to this pattern, as almost 53 per cent of the rice-grown area is irrigated.

In many water-scarce river basins of Asia, a further increase in freshwater use for agriculture would come only with a reallocation of water from other sectors. Since agriculture already consumes a considerable share of the utilizable water in these river basins, and has the lowest water use efficiency, this would face stiff resistance from other sectors of the water economy.

However, despite these constraints, the growing problems of food insecurity warrant that rice production in Asia rise significantly in the future to meet the food needs of the growing world population. It has been reported that, by the year 2025, rice production in Asia must increase by 67 per cent from the 1995 production level in order to meet the increased demand for this cereal, which is the staple for more than half of the world's population (FAORAP, 2002). This being so, the formidable challenge facing these countries, especially India, is to devise strategies for increasing the production and productivity of rice, so as to ensure the food, economic, social and water security of the region; this is argued by many scholars (Kumar et al., 2008; Kumar and van Dam, 2009).

Furthermore, water sector challenges are not just limited to its judicious use across the competing sectors. There are serious issues of governance failure in the water sector in the region, especially in India, and this failure is quite revealing in all aspects of the development and sustainable management of water resources. While Asia has some of the largest irrigation schemes in the world, the water sector in most parts of the region is beset with problems of rehabilitation and the need for modernization of irrigation schemes of varied scale. A large number of such schemes are 40 to 50 years old. In fact, many countries face serious limitations in further expansion of the irrigated area, either due to the non-availability of viable sites for construction of dams/reservoirs, problems in land acquisition for construction of canal systems, or a lack of uncommitted flows in the basins. Again, there are mammoth problems arising from lack of responsive institutional arrangements for management of the water sector. The seriousness of the issue is also evident from the fact that the PIM/IMT interventions had very limited success in most countries, including India, as is evident from a number of studies (Bassi and Kumar, Chapter 6 of this book).

Decline in public sector investment in agriculture

Despite the evidence of a strong correlation between agriculture, growth and pro-poor development outcomes in several countries in the region, there has been a significant reduction in public investment in irrigation, and a retreat by the state from promoting agriculture development during the 1980s and 1990s. Most of this retreat has been driven by factors including the launching of liberalization and structural adjustment-based economic reform policies by the national governments. The post-trade liberalization period witnessed cuts in development support by national governments and international agencies such as the World Bank, especially to developing countries. Official development assistance (ODA) for agriculture, which was quite important for improving rural infrastructure and the spread of new technology in developing countries, witnessed a sharp decline. In 2004 US$ prices, ODA declined from US$8 billion in 1984 to $ 3.4 billion by 2004 (Chand, 2008). Similarly, there has been a large decline in real lending since the late 1970s and early 1980s, when it peaked. By 1986–87, World Bank lending was only around 40 per cent of peak lending, and lending by other donors showed similar trends.

This decline in government expenditure on agricultural infrastructure in many countries was attributed to a) competing demands from non-agricultural sectors, such as health, education, social welfare and industry; b) structural adjustment programmes, which significantly reduced agricultural subsidies and other forms of farm support; and c) the declining real prices of agricultural commodities, which discouraged investment in agriculture (FAORAP, 2009, p. 11).

It is reported that annual expenditure in the agriculture sector in China and Sri Lanka was cut by nearly 50 per cent between the late 1970s and 1980s. Expenditure peaked later in Bangladesh, Indonesia and Thailand, but these countries also saw a decline in investment in irrigation (FAO, 2009). In India, public sector investment in irrigation has been either stagnant or declining since the mid 1980s (Gulati and Bathla, 2001). Data on irrigated areas, globally and across regions, show that the rate of growth in irrigated area has declined, and has been accompanied by a decline in lending for irrigation by international donors (FAO, 2009).

In contrast, rich countries continue to subsidize agriculture and protect their farmers, and their agricultural subsidies and tariffs profoundly undermine food production in developing countries. Many Asian countries saw a decline in real irrigation expenditure in the late 1980s. Ellis (2005) argues that the types of policy arrangements that were used to promote the Green Revolution in most parts of Asia – price control (fixed and floor process), buffer stocks, fertilizer and credit subsidies, public irrigation schemes, trade protection – "are largely unavailable in the current lexicon of acceptable public sector interventions". Having neglected food security and the productive sectors of their economies for several decades, many developing country governments now also lack the fiscal capacity to increase public spending in order to increase food production and agricultural productivity (Ellis, 2005).

In India, the public sector investment in agriculture and irrigation infrastructure development had grown remarkably during the 1960s through 1980s. This led to self-sufficiency in food grains. Public investment also has been quite successful in promoting private investment in the farm sector. However, the public sector's share in the total investment in the agricultural sector has declined since late 1980s. This is evident from the declining share of public sector contribution towards gross capital formation in agriculture (GCFA) in India (Figure 10.4).

Figure 10.4 shows that the share of the public sector in GCFA has declined from almost 30 per cent in 1990–91 to 18 per cent in 2008–09. One could argue that the contribution of the private sector kept on increasing, and this could have more than compensated for the decline in the contribution of the public sector. However, there are indications that the significant increase in private investment, over time, had adverse impacts.

This is because a major chunk of private investment, especially in the water sector, has gone towards groundwater development, including well drilling, the

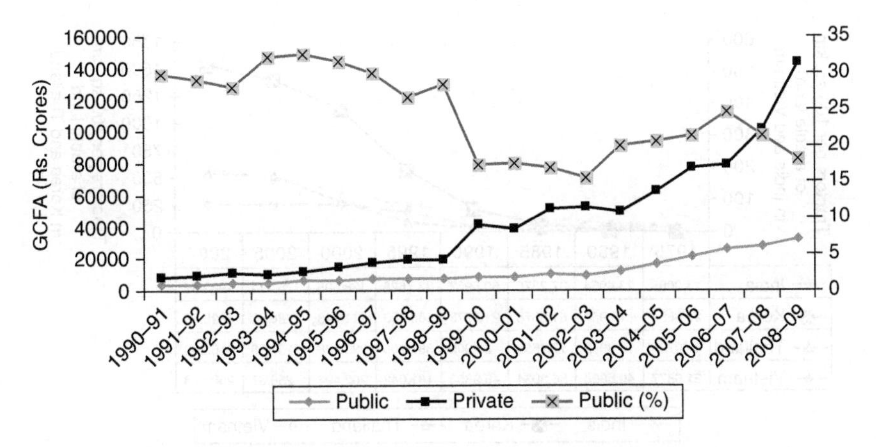

Figure 10.4 Trends in public and private sector gross capital formation in agriculture in India, 1990–91 to 2008–09.

Source: estimated based on data from the Ministry of Agriculture, Government of India.

deepening of existing wells and installation of pump sets, which ultimately caused over-extraction of groundwater resources in the semi-arid states of Punjab, Haryana, Gujarat, Rajasthan, Tamil Nadu, Madhya Pradesh and Karnataka (Kumar, 2007). Arguably, this trend has been further intensified by the populist policies of national and state governments in providing agricultural subsidies of various kinds through diverting financial resources that could have been channelized for development and strengthening of irrigation infrastructure. This is quite evident from Table 10.5, which indicates that a major chunk of agriculture subsidies in India are being provided in the form of fertilizer subsidies (45 per cent) along with power (16 per cent) and irrigation (14 per cent) subsidies, which together account for almost three-quarters of the total subsidies.

As a result of such short-sighted investment priorities, India is lagging behind many of its Asian counterparts with respect to the adoption of technologies, especially for farm mechanization – a crucial indicator of capital formation in agriculture. For instance, as shown in Figure 10.5, India's status in mechanization, as

Table 10.5 Trends in agricultural subsidies in India, 2000–01 to 2008–09

Year	*Percentage share in total agriculture subsidy*			*Total (INR crores)*
	Fertilizer	*Electricity*	*Irrigation*	
2000–01	27.2	17.6	26.1	50771 (71)[a]
2004–05	21.0	23.8	16.3	75542 (61)
2007–08	31.3	19.9	18.7	103936 (70)
2008–09	44.7	16.0	13.8	171508 (75)

Source: Ministry of Agriculture, Government of India.

Note

a. Parenthetic figures indicate the combined share of fertilizer, electricity and irrigation in total agricultural subsidies. Other subsidies include crop and food subsidies.

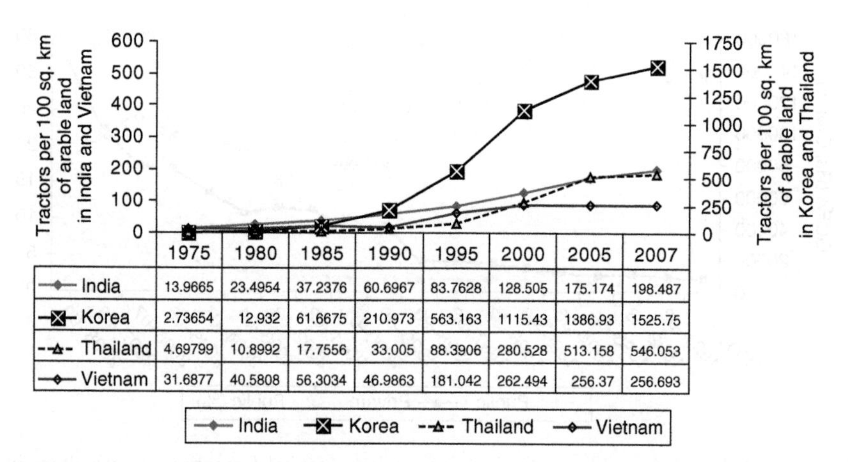

Figure 10.5 Trends in farm mechanization in India and other countries, 1975–2007.

Source: World Development Indicators dataset, World Bank, 2008.

indicated by the extent of use of tractors per square kilometre of arable, has been significantly lower than that of Korea, Thailand and Vietnam. That said, it is also important to note that, as compared to other countries, the intensity of fragmentation of farm holdings has been very high in India, thus acting as a major constraint in the large-scale adoption of agricultural technologies, including mechanized ploughing. India also has been lagging behind the other countries with respect to use of agro-chemicals, especially fertilizers. For instance, the annual average use of fertilizers in India has been as low as 1,213 kg/ha of arable land during the period 2002–07, as compared to the highest levels reported from Korea (4,536 kg/ha), Vietnam (3,743 kg/ha) and Bangladesh (1,856 kg/ha).

Of late, food security in many of the Asiatic countries, including India, has also been undermined by the variability in agro-meteorological conditions and changing natural resource availability and extraction regimes. Though many of the countries and regions are yet to experience the potential threats of the climate change events on a large scale, there are indications of climatic vulnerability becoming a reality in the years to come. The changing weather patterns can already seen to be affecting agricultural performance and the behaviour of natural environments across countries, particularly in the Asia-Pacific region. These changes are invariably associated with accelerated greenhouse gas emissions from the prevailing agricultural land use and livestock-rearing practices, inducing pressure on water supplies, speeding up desertifcation and water stress, and worsening the unpredictability and severity of weather phenomena (Wreford and Moran 2009; Golub et al., 2009).

In India and other countries in the region, the farmers' changing land use preferences, towards increased production of corn for ethanol and soybean and palm oil for biodiesel, also exert pressure on the already shrunken land area under food crops. Various studies raise concerns about the environmental sustainability of production, overall greenhouse gas emissions, and the impact on land use and food prices (IEA 2008; FAO 2008). There are strong apprehensions that, as more and more land is brought under bio-fuel crops, food prices will increase substantially, affecting poor consumers, and particularly those from low-income net food importing countries (Raju et al., 2009). Even though India is food self-sufficient, almost 50 per cent of children and practically the same number of women suffer from protein calorie malnutrition, as judged by anthropometric parameters (Bamji, 2007).

A number of empirical studies have pointed out the need to explore an array of options for sustainable production systems across countries in order to help reduce the food insecurity challenges. These options (which may be already existing or being evolved) provide avenues for a more broad-based farming/integrated systems approach to growing food and non-food crops together with the production of useful biomass, especially in marginal lands and rain-fed areas, by managing lands in a sustainable way (Vijay et al., 2009).

A number of rural innovations – technological and institutional – suggest that, if the sustainable/subsistence production systems are provided with a similar set of financial and institutional support to that provided to the conventional systems, for a sustained period of, say, 10 years, it would enhance the productivity of land and water (Down to Earth, 2007). This would eventually help reduce the food insecurity arising from the various uncertainties and challenges as described above. Such approaches to domestic food security had already been tried out by farmers through continuous innovation and adaptation in a number of regions in India and elsewhere (Pender, 2008). By and large, these innovations and adaptations remain highly localized and seem to fail to convince policy makers to change their mindset. This is reflected in the continued policy emphasis on resource-intensive, growth-oriented agriculture, irrespective of their environmental consequences and the dampening effect on food security. Unfortunately, most of the Asian economies, in particular, still continue to tinker with short-run programmes of food subsidy/public distribution of food when under political pressure to address the problem of food shortage in marginal areas and among poor rural populations.

The market integration and trade reform policies pose a new set of challenges for these countries. The reason is that emerging bilateral and multilateral free trade agreements may adversely affect the trade in food and food products. On the one hand, it raises questions about the rationale behind exporting food and food products when millions of people are deprived of the right to, and access to, food in food-exporting countries (Thailand and Vietnam in particular). On the other hand, trade agreements are increasingly determined by the ability to comply with social, environmental and health-related standards in food production. For most part of Asia, this will require investment in market infrastructure and the upgrading of farmers' technical capacity to meet the new product standards. The net result will be increased vulnerability to food insecurity, as public support systems for agriculture have been discontinued in most of these economies ever since their entry into the WTO system.

Investment strategies and technology options for sustainable agriculture: a way forward

The above discussions emphasize how imperative it is to evolve investment strategies and technology solutions to address the challenges facing sustainable development and the management of land and water resources in the Asian region. As observed, the criticality of the challenges arises from the dwindling land and water, with serious implications for food security both within and, as a large share of global food demand is met from this region, outside the region.

The need for, and the challenges involved in, ensuring sustainable investment in the development and management of land and water resources, and the technology solutions needed to achieving efficiency in farm management, are as follows.

Growing food security concerns across the world necessitate that Asian countries, in particular, sustain the dynamism of the food crops production sector that was achieved under the Green Revolution, by increasing investment in food production with a long-term perspective. In this regard, a recent study by the FAO

(2009) underscores the imperative to increase investment in revitalizing this sector. The study suggests that Asian countries need to increase total food production by 15 per cent in order to achieve domestic food security by 2015 on a priority basis (Table 10.6). Table 10.6 also reveals that countries such as Lao PDR (Laos), Bangladesh and Pakistan may have to make greater investments so as to increase the production of cereals, livestock and other food products in order to ensure domestic food security. Though the situation in India is not as critical as that of Pakistan, Lao PDR and Bangladesh, in a relative sense, it still requires sustained investment in order to increase the production of cereals and livestock (6 per cent each) in the next three year period.

The study further suggests that, in order to achieve the proposed increase in food production, Asian countries will have to bring a greater area under cultivation – 6.82 million hectares – nearly 64 per cent of which (4.32 million ha) will need to be realized through irrigation expansion. Interestingly, the study also observes that almost 43 per cent (1.87 million ha) of this area expansion should come from India. Based on these findings, it may be argued that much of the dynamism in food production in Asia depends on India's policies and development strategies. This in turn necessitates India making massive investment in bringing more land under cultivation, supported by irrigation development.

Investments for augmenting land and water resources for increased food production will be fraught with major hurdles, especially in the face of dwindling public sector investment in agriculture and deteriorating irrigation infrastructure in the region. Essentially, investments are required to improve the operation and maintenance of rice irrigation schemes through rehabilitation of the irrigation infrastructure, irrigation management transfer (IMT), irrigation modernization and a scaling up of the participatory irrigation management (PIM) programme. All these necessarily call for a reinvention of the existing irrigation management institutions, or the creation of newer ones, in order to achieve sustainable growth in agriculture along with efficiency in the allocation of water resources.

The Asian countries may face greater challenges in agricultural development if climate change-induced events adversely affect their water regimes. These

Table 10.6 Food production increments required in Asia by 2015 (per cent)

Country	*Cereals*	*Other food*	*Livestock*	*Total food*
1. Bangladesh	29	27	27	28
2. India	6	5	6	5
3. Korea	5	2	—	—
4. Viet Nam	7	3	19	9
5. Thailand	2	—	7	—
6. Pakistan	13	13	33	25
7. China	28	10	12	15
8. Lao PDR	29	27	36	30
9. Philippines	16	12	26	18
Asia	12	12	20	15

Source: FAO, 2009.

challenges become critical when we consider the fact that agricultural development in these countries already faces serious constraints due to environmental degradation and problems caused by the intensive use of farm inputs. For instance, Chand (2010) observes that, in India, the Green Revolution belt faces serious environment pollution from the burning of crop residues, as well as air, water and soil pollution caused by the use of pesticides, insecticides, weedicides and other chemicals. Similarly, Korea experienced the highest increase in greehouse gas emissions caused by a substantial increase (43 per cent) in on-farm energy consumption, along with the intensive use of fertilizer and pesticide in agriculture (OECD, 2008).

In fact, with respect to sustainable agriculture and food security, climate change throws up twin challenges to most parts of Asia, where agriculture is the mainstay of rural economy, and rice production remains the dominant agricultural activity. First, agricultural production and food systems in this vast region will have to be protected from climate-induced effects. Second, being the single largest source of greehouse gas emissions, the entire agricultural sector in these countries will have to be reoriented through climate risk adaptive cropping systems and practices to reduce the existing emission levels. Although there has been some reduction, agricultural methane emissions (CH4) still continued to contribute about 76 per cent of the greenhouse gas emissions in Thailand, followed by Bangladesh (69 per cent), Vietnam (67 per cent) and India (65 per cent) during 2005 (Viswanathan et al., 2012).

Based on the foregoing discussions and a critical review of empirical studies, it may further be stated that Asian countries, especially India, should require long-term policies and strategies for reinvigorating the agriculture sector in the emerging context of global market integration, as well as the challenges arising from internal contradictions and the potential risks of climate change. Although these countries have been largely successful in achieving food self-sufficiency through the adoption of Green Revolution technologies, major shortfalls in food supply caused by climatic aberrations, such as drought and flood, are frequent. It is reported that drought acts as the major constraint to rice production in Asia, as at least 23 million ha of rice area (20 per cent of the total rice area) are drought-prone. More importantly, the situation of India is quite precarious in this regard, as it has the largest share (59 per cent) in the total drought-prone rice area in Asia, most of which is rain-fed (Pandey et al., 2007).

It has been reported (Assaduzzaman, 2010) that, in the case of Bangladesh, the climate change impacts will require huge investments, worth US $5.5 billion per annum, for various adaptation programmes. The agriculture sector itself will require at least 50 per cent of these adaptation projects, commanding 60 per cent of the investment. Notably, adverse climate change events, such as a rise in the sea level, will have a severe effect on Bangladesh, as about 32 per cent of the total land area is in the coastal zone, with about 29 per cent of the total population living in the coastal areas (Bala and Hossain, 2010). Furthermore, over 35 per cent of farm lands in the coastal areas are affected by varying degrees of salinity and kept fallow in dry season (Karim et al., 1990).

There could be issues concerning the modus operandi required to make the flow of investment sustainable and effective for the development and maintenance of irrigation infrastructure and delivery services. In this regard, the least tried option in the Asian context is the public–private partnership (PPP). As observed by the FAO study (2009), private investment in surface irrigation systems is still in its infancy, or only contemplated in reform policy documents, mostly under pressure from external donors. A drastic shift in governance paradigm from public to the PPP model in countries like India will require a thorough overhauling of the existing (utterly ineffective) water governance regime.

More importantly, a lot more needs to be done in terms of identifying the right kind of "people-private-state" partnerships and other stakeholder involvement required for making such investments. Perhaps this issue is critically important in the wake of ever-increasing demand for water from various competing sectors, such as irrigation, livestock, drinking water, industry and ecosystem services in both rural and urban areas. In view of the mutually exclusive nature of the demand for land and water resources that emerge in many contexts, countries have to evolve long-term strategic plans involving judicious choice of appropriate PPP models for future investments in the development/management of land and water resources.

Table 10.7 provides a summary of the major areas of investment, the strategic interventions for which investments are needed, the potential sources of investment and the challenges to be addressed by Asian countries if they are to ensure sustainable agricultural development and future food security.

Conclusions

Asian agriculture is at a crossroads today, perpetuated by declining productivity growth and a declining average per capita holdings. Land fragmentation and water scarcity have become severe, triggered by population growth, rapid urbanization and industrialization. Paradoxically, rural populations depend primarily on agriculture for their livelihoods in most Asian countries, in spite of the changing structure of the economy from agrarian to services. A major fall-out of this will be the increasing vulnerability of the region's population to food insecurity and malnutrition. Strategic and policy interventions are needed to boost investment in the following areas: sustainable development and management of water resources; sustainable agriculture; and reduction of climate change impact on land and water resources. An important challenge lies in attracting investment, particularly for the modernization of old irrigation systems and farms in order to improve the productivity of land and water.

Table 10.7 The key areas for investment, the strategic interventions/policies and the key challenges

Strategic interventions	*Potential source of investment and intended outcomes*	*Challenges to be addressed*
1. Sustainable development and management of water resources		
1. Rehabilitation of old and deteriorated irrigation systems.	Public–Private Partnership with active involvement of the Farmer Cooperatives/Water Users' Associations (WUAs).	1. Huge investment requirements; 2. Weak performance of existing institutions/ineffective WUAs; 3. Private sector participation is contingent upon realization of profits. This might lead to a hike in water tariffs that will be opposed by farmers, with support from the political leadership.
2. Regulation of groundwater use.	Public–sector for development of new irrigation systems (major/minor) to reduce groundwater-intensive use and private management through adoption of appropriate market-based instruments (MBIs).	1. Difficulties in obtaining an adequate amount of land for reservoirs and canal systems. 2. Huge cost of acquiring lands in the face of growing non-agricultural demand for land as well as effective rehabilitation and resettlement of the PAP.
3. Revival of traditional water harvesting structures/local water bodies, including wetlands and development of new structures for facilitating water storage.	Public–Private Partnership with involvement of local communities and local bodies.	1. Ineffective/non-existent institutions for promoting the initiatives; 2. Lack of awareness among local communities/local bodies about the relevance of local water harnessing systems; 3. Lack of proper management systems; 4. Lack of finance for regular maintenance (desiltation and upkeep) in the case of functional management systems.
4. Large-scale promotion of water-saving technologies (WSTs)/micro-irrigation systems (MIS).	Public–Private Partnership for developing and propagating viable models of WSTs among the farmers.	1. Several physical and socio-economic constraints to adoption; 2. Absence of appropriate institutions for technology extension. 3. Inefficient water and electricity pricing and supply policies that create disincentive for adoption of WSTs. 4. Building proper irrigation and power supply infrastructure for facilitating large-scale adoption of different micro-irrigation systems.
5. Provision of drinking water in rural and urban areas.	Public–Private Partnership along with effective delivery and management systems in place based on MBIs.	1. Problems with the existing systems of water distribution in the presence of both public and private actors; poor quality of water services; pricing problems; reliability and inadequacy of the public delivery services, etc; 2. Wasteful use, without paying even the cost prices, by the more prosperous segment of consumers.

6. Inter-linking of rivers; water transfer between river basins; cleaning up of polluted rivers/ water bodies.	Public–Private partnership with proper regulations and governance systems in place through adoption of appropriate market-based instruments (MBIs).	1. Scientific challenge of assessing the utilizable flows and surplus water in different basins; 2. Engineering challenges of large-scale water transfer of water from rivers in the east to south and west; 3. Political challenges of convincing the states falling in the donor basins.
2. Sustainable agriculture		
1. Strengthening the food production systems.	Public–Private Partnership in agri-business management with proper public sector regulatory mechanisms in place to check the unscrupulous interventions of the private sector actors.	1. Lack of access to resources and infrastructure facilities, viz., land, water, credit, input markets, fair prices for produce, marketing of the final produce; 2. Infrastructure and warehouses for storage of produce; 3. Under-exploitation of the potential of agricultural produce/farm product with respect to their scope for value addition for supply in the domestic and export markets.
2. Strengthening the commercial agriculture systems, including organic agriculture and contract farming systems.	Public–Private Partnership in agri-business management with proper public sector regulatory mechanisms in place to check the unscrupulous interventions of the private sector actors.	1. Pricing and marketing problems; 2. High costs of inputs (seeds, fertilizer, pesticides, irrigation, etc) affecting profitability of farmers; 3. Increasing input use and energy intensive farming practices affecting the eco-systems; 3. Under-utilization of the potential of farm produce for value addition and supply in the domestic and export markets
3. Strengthening of fishery/ animal husbandry sectors.	Public–Private Partnership with involvement of beneficiary communities for promoting economically viable and environmentally sustainable ways of managing fishery/animal husbandry.	1. Increasing costs of management with growing market uncertainties; 2. Problems caused by diseases and natural hazards affecting the prospects of fishery and animal husbandry; 3. Depleting grazing lands along with increasing fodder costs affects profitability of animal husbandry/livestock activities.
4. Strengthening rural infrastructure, including markets, storage and institutional credit, extension systems.	Public–Private Partnership with proper regulation of the latter using MBIs.	1. Lack of rural infrastructure, such as roads and markets, affects the prospects of farming activities by depressing prices; 2. The presence of informal financial systems and increased dependency of farmers on such sources leads to indebtedness and agrarian distress.
5. Protecting the coastal agriculture systems/farm livelihood systems.	Public–Private Partnership with involvement of local communities/local bodies.	1. Increasing vulnerability of coastal communities due to changing agrarian landscapes and hydrological regimes; 2. Depletion or poor status of local resources, including common pool resources (land and water); 3. Lack of financial resources for enriching the resource base.

6. Promotion of GM technology (agri-biotechnology)/biofuel crops.	Public–Private Partnership in sharing R&D, extension and investment for developing economically viable, socially equitable and environmentally compatible (IPM/IRM) technologies and farm management practices.	1. The technological innovations are largely owned and exploited by multi national seed and agri-biotech companies without proper support/authentication by the public sector; 2. Lack of strong regulatory systems and rural institutions that facilitate an informed choice of GM crops by resource-poor farmers.
7. Gender Mainstreaming in agriculture and empowering women through institutional and technology support (as a critical strategy for achieving gender equality as per the Millennium Development Goals).	Public–Private Partnership in developing farm technologies that are specifically meant for reducing the hardships of women while engaging in agriculture/farm livelihood activities.	1. Increasing feminization of agriculture across countries in Asia, including India; 2. Compared to men, women have poor access to land and other productive assets, as well as services such as training, extension and credit; 3. Existing farm technologies are mainly designed to suit the physical constructs of male workers; 4. Significant gender disparity in wages across regions (countries/states) and agricultural activities.
3. Reducing climate change impacts affecting water and land resources		
1. Mitigation and adaptation strategies in water sector: (a) flood control and drought management; (b) IWRM including river basin management (RBM) programmes	Public–Private Partnership for making investments for adaptation and mitigation of the potential impacts of climate change.	1. Lack of financial resources for making investments for adaptation and mitigation programmes; 2. Lack of long-term perspective plans at the national and sub-national levels; 3. Lack of coordination between various governmental and non-governmental agencies engaged in water resources and agricultural development programmes; 4. Absence of an Asia-Pacific Action Plan for adaptation and mitigation programmes involving trans-boundary river basins and agro-climatic regions.
2. Mitigation and adaptation strategies in agriculture: (a) climate change resilient agriculture/farming system practices; (b) watershed development and drought proofing programmes; (c) economizing water use in agriculture.	Public–Private Partnership for investments in R&D, technologies and extension for creating and promoting the adaptation and mitigation practices.	

References

Asaduzzaman, M. (2010). *Low carbon development in Bangladesh: Agriculture and BCCSAP 2009.* Paper presented at the climate change adaptation and mitigation in agriculture science workshop, Playa del Carmen, Mexico, December 2010.

Bala, B. K., & Hossain, M. A. (2010). Food security and ecological footprint of coastal zone of Bangladesh. *Environment Development and Sustainability*, *12*(4), 531–545.

Bamji, S. M. (2007). Nutrition secure India – how do we get there? Nutrition conclave discusses the way forward. *Current Science*, *93*(11), December.

Bangladesh Bureau of Statistics (BBS). (2008). *Report on labour force survey 2005–06.* Dhaka, Bangladesh: Government of Bangladesh.

Chakravarty, S., & Dand, S. A. (2006). Food insecurity in Gujarat: A study of two rural populations. *Economic and Political Weekly*, *41*(22), 2248–2258.

Chand, R. (2008). The global food crisis: Causes, severity and outlook. *Economic and Political Weekly*, *43*(26–27), 115–123.

Chand, R. (2010). Understanding the next agricultural transition in the heartland of green revolution in India. In G. B. Thapa, P. K. Viswanathan, J. K. Routray, & M. M. Ahmad (Eds.), *Agricultural transition in Asia: Trajectories and challenges* (pp. 65–99). Bangkok, Thailand: Asian Institute of Technology.

Chung, D. K., & Dang, N. V. (2010). Agricultural transformation and policy responses in Vietnam. In G. B. Thapa, P. K. Viswanathan, J. K. Routray, & M. M. Ahmad (Eds.), *Agricultural transition in Asia: Trajectories and challenges* (pp. 145–182). Bangkok, Thailand: Asian Institute of Technology.

Down to Earth. (2007). *Crisis and opportunities in rain-fed regions.* Paper delivered at the national workshop on new paradigm for rain-fed farming: Redesigning support systems and incentives, New Delhi, India, September 27–29.

Ellis, F. (2005). *Small farms, livelihood diversification and rural-urban transitions: Strategic issues in sub-Saharan Africa.* Paper presented at the research workshop on the future of small farms, Wye College, Kent, 26–29 June. http://www.uea.ac.uk/polopoly_fs/1.53421!2005 per cent20future per cent20small per cent20farms.pdf. Accessed on 10 November 2011.

FAO. (2002). *World agriculture: towards 2015/2030. Summary Report.* Rome, Italy: Food and Agriculture Organization.

FAO. (2008). Biofuels: Prospects, risks and opportunities services. In FAO, *The state of food and agriculture 2008.* Rome, Italy: Food and Agricultural Organization.

FAO. (2008a). *The state of food insecurity in the world 2008.* Rome, Italy: Food and Agriculture Organization.

FAO. (2009). *The state of food and agriculture, 2009.* Rome, Italy: Food and Agriculture Organization.

FAO. (2010). *Agricultural biotechnologies in developing countries. Options and opportunities in crops, forestry, livestock, fisheries and agro-industry to face the challenges of food insecurity and climate change: Current status and options for crop biotechnologies in developing countries (ABDC-10).* Paper delivered at the FAO international technical conference, Guadalajara, Mexico, 1–4 March. www.fao.org. Accessed on 4 November 2011.

FAORAP (Food and Agriculture Organization Regional Office for Asia and the Pacific). (2002). *Investment in land and water: Proceedings of the regional consultation, Bangkok, Thailand, 3–5 October 2001.* Bangkok, Thailand: RAP Publications.

FAORAP. (2006). *Rapid growth of selected Asian economies: Lessons and implications for agriculture and food security, Republic of Korea, Thailand and Vietnam.* Bangkok, Thailand: RAP Publications. ftp://ftp.fao.org/docrep/fao/009/ag088e. Accessed on 10 November 2011.

FAORAP. (2009). *Selected indicators of food and agricultural development in the Asia Pacific region 1998–2008*. Bangkok, Thailand: RAP Publications.

Golub, A., Hertel, T., Lee, H-L., Rose, S., & Sohngen, B. (2009). The opportunity cost of land use and the global potential for greenhouse gas mitigation in agriculture and forestry. *Resource and Energy Economics*, *31*, 299–319.

Gulati, A., & Bathla, S. (2001). Capital formation in Indian agriculture: Re-visiting the debate. *Economic and Political Weekly*, *36*(20), 1697–1708.

Hirway, I., & Mahadevia, D. (2003). *Status of health and nutrition in Gujarat: Some issues*. Observer Research Foundation Working Paper. New Delhi, India: Observer Research Foundation.

IEA. (2008). *World energy outlook*. Paris, France: International Energy Agency Publications. http://www.fas.usda.gov. Accessed June 2010.

Kaosa-ard, M. S., & Rerkasem, B. (2000). *The growth and sustainability of agriculture in Asia: A study of rural Asia* (Vol 2). Hong Kong: Oxford University Press.

Karim, Z., Ibrahim A., Iqbal, A., & Ahmed, M. (1990). *Drought in Bangladesh: Agriculture and irrigation schedules for major crops*. Dhaka, Bangladesh: Bangladesh Agricultural Research Council (BARC).

Keller, J. (2002). *Evolution of drip/micro-irrigation: traditional and non-traditional uses*. Keynote paper at the international meeting on advances in drip/micro-irrigation, Puerto de la Cruz, Tenerife, 2–5 December.

Kumar, M. D. (2007). *Groundwater management in India: Physical, institutional and policy alternatives*. New Delhi, India: Sage Publications.

Kumar, M. D., Turral, H., Sharma, B., Amarasinghe, U., & Singh, O. P. (2008b). Water saving and yield enhancing micro irrigation technologies: When do they become best bet technologies? In M. D. Kumar (Ed.), *Managing water in the face of growing scarcity, inequity and declining returns: Exploring fresh approaches*. Proceedings of the 7th annual partners' meet, IWMI-Tata Water Policy Research Program, ICRISAT, Patancheru, 2–4 April, 2008.

Kumar, M. D., & van Dam, J. (2009). Improving water productivity in agriculture in developing economies: In search of new avenues. In M. D. Kumar, & U. Amarasinghe (Eds.), *Water productivity improvements in Indian agriculture: Potentials, constraints and prospects*. Strategic analysis of national river linking project of India series 4. Colombo, Sri Lanka: International Water Management Institute.

Kumar, M. D., Scott, C. A., & Singh, O. P. (2011). Inducing the shift from flat rate or free agricultural power to metered supply: Implications of groundwater depletion and power sector viability in India. *Journal of Hydrology*, *409*, 382–394.

Lee, S-W. and Kim, H. J. (2010). Agricultural transition and rural tourism in Korea: Experiences of the last forty years. In G. B. Thapa, P. K. Viswanathan, J. K. Routray, & M. M. Ahmad (Eds.), *Agricultural transition in Asia: Trajectories and challenges* (pp. 37–64). Bangkok, Thailand: Asian Institute of Technology.

Meinzen-Dick, R. S, Quisumbing, A., Behrman, J., Biermayr-Jenzano, P., Wilde, V., Noordeloos, M., Ragasa, C., & Beintema, N. (2011). *Engendering agricultural research, development, and extension*. Research monograph. Washington, DC, USA: International Food Policy Research Institute.

Mittal, S., & Deepthi, S. (2009). *Food security in South Asia: Issues and opportunities*. Working Paper 240. New Delhi, India: Indian Council for Research on International Economic Relations (ICRIER).

Mukherji, A., Facon, T., Burke, J., de Fraiture, C., Faurès, J.-M., Füleki, B., Giordano, M., Molden, D., & Shah, T. (2009). *Revitalizing Asia's irrigation to sustainably meet tomorrow's food needs*. Colombo, Sri Lanka: International Water Management Institute.

OECD. (2008). *Environmental performance of agriculture in OECD countries since 1990*. Paris, France: OECD.

Pandey, S., Bhandari, H., & Hardy, B. (Eds.). (2007). *Economic costs of drought and rice farmers' coping mechanisms: A cross country comparative analysis*. Manila, Philippines: International Rice Research Institute.

Patnaik, U. (2009). Origins of the food crisis in India and developing countries. *Monthly Review*, JulyAugust.

Pender, J. (2008). *Agricultural technology choices for poor farmers in less-favoured areas of South and East Asia*. Rome, Italy: International Fund for Agricultural Development (IFAD).

Phrek, G. (2010). Drivers of change in Thai agriculture: Implications for development and policy in the new millennium. In G. B. Thapa, P. K. Viswanathan, J. K. Routray, & M. M. Ahmad (Eds.), *Agricultural transition in Asia: Trajectories and challenges* (pp. 100–144). Bangkok, Thailand: Asian Institute of Technology.

Pray, C. E., Ramaswami, B., & Kelley, T. (1998). *Liberalization, private plant breeding and farmers' yields in the semi-arid tropics of India*. Unpublished paper. Washington, DC, USA: The World Bank.

Raju, S. S., Shinoj, P., & Joshi, P. K. (2009). Sustainable development of biofuels: Prospects and challenges. *Economic and Political Weekly*, *40*(52), 65–72.

Rawal, V. (2008). Ownership holdings of land in rural India: Putting the record straight. *Economic and Political Weekly*, *43*(10), 43–47.

Song, J. H. (2006). Perspectives on agricultural development in the Republic of Korea: Lessons and challenges. In FAORAP (Ed.), *Rapid growth of selected Asian economies: Lessons and implications for agriculture and food security, Republic of Korea, Thailand and Vietnam* (pp. 1–23). Bangkok, Thailand: RAP Publications.

Thapa, G. B., Viswanathan, P. K., Routray, J. K., & Ahmad, M. M. (Eds.) (2010). *Agricultural transition in Asia: Trajectories and challenges*. Bangkok, Thailand: Asian Institute of Technology.

Vijay, K. T., Raidu, D. V., Killi, J., Pillai, M., Shah, P., Kalavadonda, V., & Lakhey, S. (2009). *Ecologically sound, economically viable: Community managed sustainable agriculture in Andhra Pradesh, India*. Washington, DC, USA: The World Bank.

Viswanathan, P. K., Thapa, G. B., Routray, J. K., Ahmad, M. M. (2012). Agrarian transition and emerging challenges in Asian agriculture: A critical assessment. *Economic and Political Weekly*, *47*(4), 41–50.

Woolverton, A., Regmi, A., Tutwiler, M. A. (2010). *The political economy of trade and food security*. Geneva, Switzerland: International Centre for Trade and Sustainable Development.

World Bank, The (2008). *World development report 2008: Agriculture for development*. Washington, DC, USA: The World Bank.

Wreford, A. & Moran, D. (2009). *Climate change and agriculture: Impacts, adaptation and mitigation*. Paper presented at the climate smart food conference, Lund, Sweden, 23–24 November. http://www.se2009.eu/polopolyfs/1.24463! menu/standard/file/Anita per cent20Wreford.pdf. Accessed on 8 September 2010.

11 Water management for food security and sustainable agriculture

Strategic lessons for developing economies

M. Dinesh Kumar

Introduction

Sub-Saharan Africa is largely agrarian. Yet the region has the unique distinction of having the lowest level of cereal yields and agricultural productivity growth rates (Rosegrant et al., 2001; von Braun, 2007; FAO, 2006). The region is the most water-stressed in the world (UN HDR, 2006). Yet only a small fraction of the utilizable water resources of the region have so far been tapped (Falkenmark and Rockström, 2004). In the absence of other economic opportunities in rural areas, poverty reduction is closely linked to water development for irrigated agriculture (FAO, 2007). However, the region is yet to see significant investment in water resources development, including irrigation development (Rosegrant et al., 2001). The region suffers from inadequate human resource capacities in the water sector, as well as from poor finances (Falkenmark and Rockström, 2004). It is, thus, also the most food insecure region in the world (IFPRI, 2011; Weismann, 2006), depending largely on donor aid and food imports (von Braun, 2007).

In the 1960s, most of the Asian continent, barring the Far Eastern economic giants, looked the same way sub-Saharan Africa looks today, as far as agriculture, food security and poverty is concerned. But Asian countries, including India and China, made significant strides in terms of maintaining high growth in agricultural productivity and production, lifting hundreds of millions of people out of poverty. Irrigation development has played a crucial role in rural poverty reduction. The region has moved away from a "low resource use/low productivity" regime to, largely, an "intensive resource use/moderate to high productivity" regime. While, in the semi-arid and arid parts of Asia, groundwater has played a significant role in revolutionizing irrigation, the latter has also resulted in the problems of "aquifer mining", exacerbated by a lack of adequate planning, legal framework and governance; and this has threatened the long-term sustainability of irrigated agriculture (FAO, 2007).

Sub-Saharan Africa, which is riddled by conflicts, political instability, poor governance, corruption, politics of exclusion, high rural poverty (Ong'ayo, 2008) and weak human resource capacities (Falkenmark and Rockström, 2004) is nowhere comparable with India. Yet there are regions within India which are as poor and food insecure as some countries of sub-Saharan Africa. Because of the heterogeneous agro-climate and socio-economic conditions prevailing across the country, the Indian experience provides important lessons for sub-Saharan Africa, on such issues as what the the long-term strategies should be for food security and poverty reduction through the agricultural growth route; and what can possibly go wrong. This chapter synthesizes the major findings from all the previous chapters, and draws lessons for sub-Saharan Africa. It also discusses some key areas for future research that would contribute to framing national policies for food security.

Lessons from India for sustainable agricultural production

Policies should work in harmony and not at cross purposes

In India, several policies relating to agricultural production work at cross purposes. Let us take the example of micro-irrigation systems. The government of India has been providing subsidies for the promotion of micro-irrigation systems, particularly drips and sprinklers, taking the view that the use of these technologies helps save scarce water resources, enhances crop yields and saves labour. But the water saving benefit is a social benefit. It does not lead to a private benefit for the farmers in many situations. This is because the farmers are neither confronted with the marginal cost of using water or electricity (due to an absence of volumetric pricing of water or pro-rata pricing of electricity), nor the opportunity costs of using these resources (Kumar and Singh, 2001; Kumar, 2005; Kumar et al., 2011). But, while governments show great enthusiasm to promote micro-irrigation systems, there is hardly any political will to fix water and energy prices in the farm sector, which would actually provide the farmers with an increased incentive to use both resources efficiently.

The large scale adoption of micro-irrigation systems in some of the water-scarce regions in India is mainly because that they are the best bet technology to irrigate some of the high valued crops, in order to obtain high yields. Another reason is that there is a physical shortage of irrigation water in wells due to the excessive depletion of groundwater. In other words, had the state governments in the water-scarce areas of the country shown the political will to address water and energy pricing, the level of adoption of micro-irrigation systems and many of the water-efficient irrigation practices[1] in these regions would be far higher than those existing at present.

Another issue is of power supply and irrigation water delivery policies. In many states, the power supply schedules followed in the farm sector does not encourage the efficient use of micro-irrigation systems, which are mostly energy intensive and which demand the daily watering of crops. Since the revenue generated through the supply of electricity in the farm sector is very low, owing to heavy subsidy, the tendency is to provide poor quality, cheap power. This leaves little room for farmers to schedule irrigation and deliver water in a controlled manner. For farmers to use micro-irrigation systems under such conditions, large intermediate storage systems become essential. In contrast, flood irrigation is most amenable to such supply regimes, because a large amount of water is stored in the soil profile in order to meet 5–7 days of crop water requirements.

As regards water delivery policies in canal irrigation, in most irrigation schemes, water is supplied through gravity systems. For farmers to use micro-irrigation systems in their plots, these are several pre-requisites. First, an intermediate storage system would be required for storing the water, which comes at a frequency of 7–14 days. Second, a pumping device is needed for lifting the water from the storage and for pressurizing the MI system. The cost of building this additional infrastructure would obviously reduce the economic benefit from such systems. The only place in India where farmers have used micro-irrigation systems in the canal command is Rajasthan. The government here had provided huge subsidies for building a storage system called Diggie, in addition to subsidizing micro-irrigation systems.

Another important issue is of the potential impact of introducing micro-irrigation on domestic and regional food security. The fact that MI systems either support or become the best bet technologies for high valued cash crops like cotton, groundnut, potato, fruits, sugarcane and vegetables, but not so much for cereals like wheat, bajra and paddy, should be kept in mind while designing policies for their promotion. In those regions which are large contributors to national cereal production, care should be exercised when framing agricultural policies, to ensure that the adoption of MI systems and crops which are amenable to them do not expand at the cost of traditional cereals. Farmers should be provided with sufficient motivation to produce cereals, through proper market support and monetary incentives.

Micro level water harvesting should be supported by macro planning

Over the past two decades, small and decentralized water harvesting has been at the forefront of the water management debate in India. It has been put forward as a solution for water problems in many water-scarce regions and is included in various programmes including those for watershed development in rain-fed areas (Kumar et al. 2006; Kumar et al., 2008; Ray and Bijarnia, 2006; Syme et al., 2011). This has been the result of initiatives by NGOs to popularize small water harvesting systems as community-based initiatives to address local water problems. Decentralization, in the complete absence of scientific and technical manpower at the level of local Panchayats, itself meant there was a lack of coordinated and scientific planning, and poor technical supervision over execution – resulting in ineffective schemes and poor quality structures. There has been no serious consideration of catchment hydrology, topography and geo-hydrology in deciding the number and size of structures and their location in the catchment. This has led to ecological degradation of catchments/basins due to the over-appropriation of surface water, as well as several negative downstream impacts (Batchelor et al., 2002; Kumar et al., 2008; Ray and Bijarnia, 2006). More importantly, the decision to implement small water harvesting schemes is not driven by considerations of irrigation and other water benefits, but employment generation (Bassi and Kumar, 2011). As “scale effects” are not considered (Kumar et al., 2006; Syme et al., 2011), this leads to over-doing it. If long-term benefits are to be derived from such schemes, they need to be based on scientific planning which considers basin and catchment hydrology, sound engineering, economics and ecology.

Small water harnessing systems, such as tanks and ponds, built in naturally water-scarce regions of India have poor dependability, due to a high variability in rainfall and insufficient catchments. While, in good rainfall years, they might be able to meet the multiple water needs of the local communities, in bad rainfall years many of the water uses would seriously suffer, and water might get diverted by the influential groups within the villages for more commercial needs. Wherever the possibility exists, these small water harnessing systems will have to be integrated with large water systems. Such systems, also known as "wine and melon" systems, already exist in the commands of large irrigation systems. Water from irrigation canals is fed into small ponds and tanks in the command at the time of excess inflow from their catchments, and used by the communities when the release from the large system dwindles.

Policies should be designed at the disaggregated level

The design of government subsidies for micro-irrigation systems is not based on a sound understanding of whether welfare benefits are accrued from their adoption or not, and, if so, what kind of benefits. There are many issues involved in this regard. First of all, the impacts, such as water saving and labour saving, do not translate into welfare gains everywhere. In a region where water is abundant, saving water through MI systems does not lead to any welfare gain. Similarly, labour saving does not result in welfare gains everywhere. There are agriculturally prosperous regions which also experience an acute shortage of labour in agriculture. If MI systems are adopted in such regions, labour saving would lead to societal benefits, in addition to (input) cost saving benefits for the adopter farmers. But, the same will not be true for regions where agricultural labour is in surplus and labourers do not get paid adequately. Bihar is one example. Large scale adoption of MI systems in such regions would lead to greater hardship for farm labourers.

It is a misnomer that micro-irrigation systems uniformly save water, irrespective of the crops, climate, physical environment and the type of technology used. The real water saving benefits of micro-irrigation depend on the type of MI technology, the type of crop for which the system is used, the type of soils, the geo-hydrological environment and the climate. The systems are likely to produce the intended benefit of real water saving if used for distantly-spaced crops, under arid and semi-arid climatic conditions and under deep water table conditions. The reason is that only under such conditions would the system help reduce the non-beneficial evaporation and non-recoverable deep percolation of water (Kumar et al, 2008; Kumar, 2009). On the other hand, use of MI systems can cause negative welfare effects in certain situations. For instance, in regions where the groundwater table is shallow, the conventional method of irrigation will result in recharge to groundwater from return flows from irrigated fields and the improved sustainability of well irrigation. Here, adoption of micro-irrigation may not result in any significant water saving, but will have negative third party effects (Dhawan, 2000). Hence, it would not be appropriate to provide subsidies for use of MI systems.

However, these aspects are not considered in the design of subsidy policies. Central subsidies are available for all states, whereas state subsidies might vary. It would be inappropriate to provide subsidies for MI systems if the real welfare benefits are not realized through their use.

There is an abundance of groundwater in eastern India, and its use is very low for agricultural production. In such regions, the power pricing policies should be such as to encourage greater use of groundwater by the farmers. A flat rate system of pricing electricity will be most appropriate there. In order to ensure greater equity in the distribution of welfare benefits, it is important to provide subsidies for well drilling and power connections, as only a small fraction of the farmers there own wells. In sum, an electricity pricing policy that is uniform across the country will be untenable when there is high degree of variation in groundwater resource endowment.

Broaden the objectives and criteria for assessing the performance of large water systems

It is not just too simplistic, but dangerous, to assume that all large dams cause huge social and environmental problems (Biswas and Tortajada, 2001). First of all, the criteria for defining water systems as "large" and "small" have to be broadened, so as to meet the objective of reflecting the real negative social and environmental consequences of building large water systems, and not the objective of reflecting the engineering challenges in building the dams (Shah and Kumar, 2008). Currently dam height and storage volume are used as norms for classifying dams as large. But empirical analysis, based on a global database on large dams, shows that dam height and storage volume hardly indicate the social and environmental problems associated with large dams. If the objective is to assess the ecological damage and the number of people displaced by reservoirs, the criteria for assessment will need to be the area submerged or the reservoir area.

Furthermore, carrying out a benefit–cost analysis of large water resource projects on the basis of the simple criteria of a) incremental income from the area directly irrigated by the canals; b) revenue generated from the hydropower generated; and c) the number of people served by the drinking water supplies, would be highly misleading (Shah and Kumar, 2008; and also Chapter 4 of this book). As indicated by analysis of data from the canal command area of the Sardar Sarovar Narmada project, there are huge indirect benefits which are hard to foresee at the time of project planning. They can change the benefit–cost equations. The economic value of the indirect benefits, such as employment generation, increased wage rates for agricultural labourers, improved sustainability of well irrigation, savings in the economic cost of the energy used for groundwater pumping, and improved sustainability of drinking water wells, etc., can sometimes be larger than the direct economic benefits.

When semi-arid and arid regions across the world are facing problems of groundwater over-exploitation (Kumar, 2007), improved recharge to groundwater through the introduction of canal irrigation in such regions comes as a boon. While large canal irrigation projects faced criticism for the environmental problems of water-logging and salinity in the command areas (Shah and Kumar, 2008), under changed circumstances they, in reality, become a solution to the larger environmental problems of groundwater mining (Vyas, 2001; Shah and Kumar, 2008). Surface water importation as a solution for groundwater mining problems assumes greater importance in light of the fact that local water harvesting and groundwater recharging interventions are not going to be effective in naturally water-scarce regions.

Food insecurity in sub-Saharan Africa and its supply dimensions

Where does sub-Saharan Africa stand in terms of food security?

Most of the sub-Saharan African countries are highly food insecure. This is evident from the 2006 Global Hunger Index scores for 118 countries (Weismann, 2006) which takes into account the percentage of people with under nourishments, the child mortality rate and the percentage of underweight children in these countries. According to the latest Global Hunger Index data published by IFPRI, out of the 26 countries which have alarming to extremely alarming hunger index scores, 22 are in the African continent (IFPRI, 2011).

An analysis of the progress in reducing hunger, expressed in terms of a reduction in GHI scores among countries, shows that progress has been relatively less in sub-Saharan Africa than in many countries in South East Asia and Latin America. The reduction in GHI scores ranges from 0.0–24.9 per cent for some in southern and central Africa, to 25.0–49.9 per cent for some others. A few countries in sub-Saharan Africa showed an increase in hunger. At the same time, the reduction in the GHI score has been much higher (above 50 per cent) for many countries of Latin America (Brazil, Uruguay, Chile), the Middle East, Central Asia (Turkey) and China. The achievement of India in reducing hunger was less than that of more populous countries such as China. The GHI score for the country went down from 30.4 per cent to 23.7 per cent during the period from 1990 to 2011.

What is even more alarming is the fact that both the percentage and aggregate number of undernourished children increased in sub-Saharan Africa during the period from 1992–04 (von Braun, 2007, Figure 13), and is expected to increase until 2020, as per the IFPRI forecasts made in 2001: to 39 million from 33 million in 1997 (based on Rosegrant et al., 2001, Figure 12). The little progress in reducing hunger in sub-Saharan Africa has come from a reduction in the percentage of undernourished people in that region.

The root cause of food insecurity in sub-Saharan Africa

One of the reasons for food insecurity in sub-Saharan Africa is poor agricultural growth. Cereal yields in sub-Saharan Africa are the lowest amongst all the regions of the world. During the three decades from 1967 to 1997, the increase in cereal yields has been negligible, whereas, during the same period, yields doubled in South Asia 2,000kg/ha (Rosegrant et al., 2001, based on FAOSTAT).

The impediments to agricultural growth are many, and there is a complexweb of problems. But some of the most critical ones are: a) poor investment in irrigation; b) poor adoption of modern agricultural technologies, including high yielding varieties and farm machinery; c) poor workforce in agriculture; d) poor extension services in the agricultural sector; and e) poor market infrastructure (FAO, 2009). These are problems on the socio-economic front. These problems are compounded by high rainfall variability and the frequent occurrence of droughts, though variability in rainfall is high in the drier regions, and low in the wetter regions (Gommes and Petrasi, 1994). On the governance front, there are weak institutions, a non-vibrant civil society, a weak agricultural research system characterized by an inadequate human resource base, and rising food prices. Out of the 24 countries which have low human development indices (those which are less than 0.50), 23 are in Africa (UN HDR, 2009). Many of these problems can be averted through improvements in the water security of the people in this region.[2]

Worldwide, experiences show that improved water security (in terms of access to water; levels of use of water; the overall health of the water environment; and enhancement of the technological and institutional capacities needed to deal with the sectoral challenges) leads to better human health and environmental sanitation; food security and nutrition; livelihoods; and greater access to education for the poor (see, for instance, UNDP, 2006). This aggregate impact can be segregated, with irrigation having a direct impact on rural poverty (Bhattarai and Narayanamoorthy, 2003; Hussain and Hanjra, 2003), food security, livelihoods, nutrition (Hussain and Hanjra, 2003) and the number of people in the productive workforce; and domestic water security having positive effects on health and environmental sanitation, with spin-off effects on livelihoods and nutrition (positive), school dropout rates (negative) and the productive workforce.

Currently, the accessibility of safe water is very scant in sub-Saharan Africa, with only 22 to 34 per cent of the populations in at least eight sub-Saharan countries having access to safe water. UNEP projects that, in the year 2025, as many

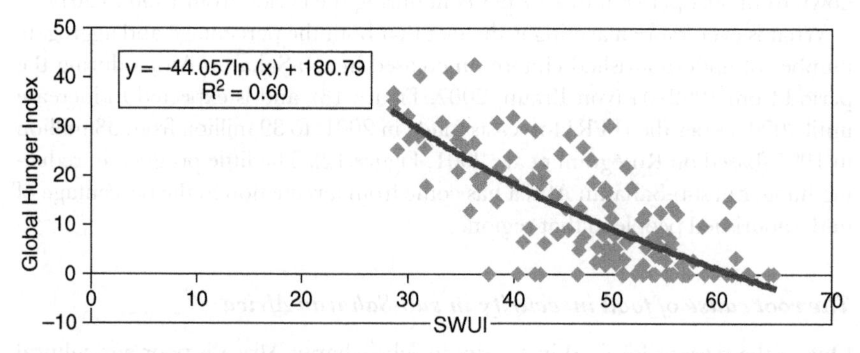

Figure 11.1 Global Hunger Index vs. SWUI.

Source: authors' own analysis of the secondary data.

as 25 African nations – roughly half the continent's countries – are expected to suffer from a greater combination of increased water scarcity and water stress. Dirty water and poor sanitation account for the vast majority of the 0.8 million child deaths each year from diarrhoea, making it the second largest cause of child mortality. According to the UN Human Development Report (2006), diseases and productivity losses linked to water and sanitation amount to 5 per cent of GDP in sub-Saharan Africa. Women bear the brunt of responsibility for collecting water, often spending up to 4 hours a day walking, waiting in queues and carrying water; water insecurity linked to climate change threatens to increase malnutrition to between 75 and 125 million people by 2080, with staple food production in many sub-Saharan African countries falling by more than 25 per cent (UN HDR, 2006, p. 37).

The strong inverse relationship between the Sustainable Water Use Index, which captures the overall water situation in a country (Kumar et al., 2008; Kumar, 2009), and the Global Hunger Index (GHI), developed by IFPRI for 118 countries (Weismann, 2006), provides broader empirical support for some of the phenomena discussed above. In addition to these 118 countries for which data on GHI are available, we have included 18 developed countries. For these countries, we have taken zero values, assuming that these countries do not face problems of hunger. The estimated R^2 value for the regression between SWUI and GHI is 0.60. The coefficient is also significant at 1.0 per cent level. It shows that, with an improved water situation, the incidences of infant mortality (those below 5 years of age) and impoverishment reduce. In that case, an improved water situation should improve the value of the human development index, which captures three key spheres of human development such as health, education and income status (Kumar, 2009).

Therefore, the root cause of the problems of food insecurity in sub-Saharan Africa lies in water insecurity (Kumar et al., 2008; Shah and Kumar, 2008). There have been recent attempts to link food insecurity to national income using the correlation between per capita GDP and food insecurity using time series data for different regions (IFPRI, 2011; Weismann, 2006). This is mainly because of auto-correlation, i.e., countries having high water security (expressed by us in terms of SWUI) also have high per capita income. The fact, as illustrated by Kumar et al. (2008) and Shah and Kumar (2008), is that water security remains, for many countries, crucial to achieving progress in the form of high human development indicators and economic growth. Water security in these, semi-arid to arid tropical countries will come only through water development (Falkenmark and Rockström, 2004), either through the building of storage or through judicious exploitation of groundwater.

Challenges in achieving food security

Sub-Saharan Africa has less than 25 per cent of its cultivable land under crop production (FAO, 2009). The region has the lowest irrigated to rain-fed area ratio, at less than 3 per cent (FAO, 2006, Figure 5.2, p. 177). The drought-prone areas of sub-Saharan Africa are characterized by one of the lowest levels of agricultural productivity in the world, primarily due to water stress during crop growth. One of the reasons for the poor utilization of arable land for cultivation is the lack of certainty in obtaining yields, in the wake of uncertain rainfalls and the absence of irrigation facilities.

Even the economic growth of this predominantly agrarian region is closely correlated with rainfall (Barrios et al., 2004; Foster et al., 2006). Irrigation is the key to improving water security, expanding crop land and also raising crop yields in the drought-prone areas. But one of the biggest challenges in reducing the region's vulnerability to droughts and its associated problems of food insecurity and hunger is in developing irrigation (FAO, 2009; FAO, 2007). Currently, only two of the sub–Saharan African countries have extensive irrigation. They are South Africa and Madagascar, with 1.43 million ha and 1.15 million ha of irrigated area respectively (You et al., 2007).

As we have seen above, in addition to irrigation, improved access to safe water and improved sanitation will also go a long way in reducing child mortality through the control of fatal water-borne diseases. It will help improve education by reducing school dropout rates. It will increase the productive workforce by improving family health and nutrition. But, this is beyond the scope of this chapter, and here we will only deal with water management for agricultural and food production.

In India, the initial impetus in irrigation development came from the public sector, and mainly covered large, medium and minor surface irrigation systems (Shah, 2009). But, later on, with an advancement in drilling technology, massive rural electrification, the institutional financing of well development, and heavy subsidies for electricity for agricultural use, groundwater development for irrigation took off in the countryside, with well irrigation becoming intensive in the semi-arid and arid parts of the country (Kumar, 2007). Today, well irrigation surpasses surface irrigation and accounts for nearly two-thirds of the net irrigated area in India (Kumar, 2007; Shah, 2009).

But there is little reason to believe that the irrigation development trajectory would be more or less the same in sub-Saharan Africa, given the drastically different socio-political scenario and human resource capacities of the countries of that region. Surface irrigation development in sub-Saharan Africa is likely to happen at a slow pace even in the coming years, and too little can be done about changing it from a water sector perspective, unless the macro level issues of political instability, corruption in governments, institutional capacity and finance are addressed too. These are in addition to the host of social and environmental issues which the surface water projects raise. Corruption in government is likely to reduce donor confidence in many of the sub-Saharan countries.

Most of the leverage, therefore, lies in groundwater development through private sector initiatives. But, unlike India, where drilling technology has come very handy in rural areas due to low cost and easy accessibility, well drilling is very expensive in most African countries. Apart from high drilling costs, there are also technical issues. They concern uncertainty about groundwater resource conditions. As noted by Foster et al. (2006), the scientific information about groundwater resource conditions in sub-Saharan Africa is patchy at the local level, while some information about aquifer characteristics, recharge and abstraction are available at the regional level. Unless farmers are convinced of their ability to hit water, investment is unlikely to come, even if funds are available. An associated challenge is in rural electrification.

A further challenge is to change the investment climate. In addition to irrigation, there is a need for greater investment in public goods that support agriculture, such as research and extension, rural roads, storage facilities, education and health (FAO, 2009).

Land and water management strategies for sustainable agriculture production and food security

Planning small and large surface water systems in river basins in integrated manner

In the early stages of irrigation development in independent India, the focus was on large irrigation systems involving reservoirs and barrages for water storage and diversion, and canal systems for water distribution and delivery. The work of building irrigation schemes was mainly in the domain of public enterprises, principally the irrigation departments of various provinces. The 1970s, 1980s and 1990s saw expansion in groundwater irrigation, which occurred mostly with private investment from farmers. The past two decades have, however, seen an increasing focus on small water harnessing systems, with an accent on decentralization and community participation. This has emerged in response to growing water scarcity in semi-arid and arid regions, with the supplies from both public irrigation systems and groundwater-based sources being unable to meet the growing demand for water in agriculture. The underlying assumption, perhaps, was that small structures could be managed easily through local efforts, without much help from public agencies whose performance in managing large irrigation schemes has, in any case, not been impressive.

But these efforts have not produced any laudable results, due to lack of scientific planning and over-doing it, without any attention being paid to basin hydrology. In fact, most of the structures were built in basins which already have a sufficient number of water systems built to appropriate its limited water resources. The scale effects of small scale water harvesting and watershed development at the macro level were ignored (Kumar et al., 2006; Kumar et al., 2008; Syme et al., 2011). This has led to the situation of dividing the waters within the basin rather than augmenting the utilizable water resources. This had seriously impaired the economic viability of the structures, apart from causing the ecological problem of reduced streamflows in downstream parts (Kumar et al., 2006; Kumar et al., 2008). This is leading to a phenomenon that Falkenmark and Rockstrom refer to as "over-crowding" in water, and technical water stress.

Many African countries are also caught up in the movement for small water harvesting. There has been some work done in eastern African countries (Oweis and Kijne, 1999; Rockström et al., 2002), Mexico (Scott and Silva-Ochoa, 2001) and India to show the impact of water harvesting on crop water productivity. Rockström et al. (2002) have shown the remarkable effect of supplementary irrigation through water harvesting on the physical productivity of water, expressed in kg/ET, for crops such as sorghum and maize. However, the research did not evaluate the incremental economic returns, due to the supplementary irrigation, against the incremental costs of water harvesting. It also does not quantify the real hydrological opportunities available for water harvesting at the farm level, nor its reliability. The work by Scott and Silva-Ochoa (2001) in the Lerma-Chapala basin in Mexico showed higher gross value product from crop production in those areas with a better allocation of water from water harvesting irrigation systems. But figures of surplus value product, which take into account the cost of irrigation, were not available from their analysis. In arid and semi-arid regions, the hydrological and economic opportunities of water harvesting are often over-played (Kumar and van Dam, 2009). A recent piece of work in India has shown that the cost of water harvesting systems would be enormous, and reliability of supplies from it very poor in arid and semi-arid regions, which are characterized by low mean annual rainfalls, very few rainy days, high inter-annual variability in rainfall and rainy days, and high potential evaporation, leading to a much higher variability in runoff between good rainfall years and poor rainfall years (Kumar et al., 2006). These findings are applicable to sub-Saharan Africa, as most of it experiences high variability in rainfall and droughts.

With the high capital cost of water harvesting systems, small and marginal farmers would have less incentive to use them for supplementary irrigation, as incremental returns due to yield benefits may not exceed the cost of the system. This is particularly so for crops having a low economic value (Kumar and van Dam, 2009), which dominate cropping in sub-Saharan Africa. The dominant crops of the region are cassava, maize, millet, cow pea, barley (all food grains), bean (vegetable) and cotton, groundnut, cocoa and coffee. Among these crops, only a few of the high valued ones-require irrigation. The other crops, such as oil palm, banana, sugarcane, potatoes, yam, rice and wheat are grown in a small area (You et al., 2007).

Since sub-Saharan Africa is still at the early stage of water development, there is a scope for using an integrated catchment approach in thebasin-wide planning of small and large water resource systems, in order to avoid the phenomenon that Falkenmark and Rockstrom refer to as "techno-economic water scarcity", wherein the number of people competing for a unit of water is excessively large. Such tasks should be entrusted with a competent scientific/technical agency outside the jurisdiction of public irrigation enterprises. This would enable sufficient scientific input into planning the water systems, big or small. Small and large systems have to co-exist. Water resource development should begin with small structures. The small water harvesting systems should be built in an area only if the local hydrology and topography permit, and decisions to plan and build such structures should not be driven by mere political considerations. The large water systems can be planned downstream of the small structures, in order to harness the untapped (surplus) water from these local catchments.

Careful and scientific planning and development of Africa's groundwater

Available data on groundwater development in African countries shows a very low degree of groundwater development at the regional scale in sub-Saharan Africa (Foster et al., 2006; Siebert et al., 2010). But planning for groundwater development cannot be driven by macro level information about groundwater recharge and abstraction. Micro level information about utilizable resource and abstraction and the stage of development would be essential. Furthermore, given the complex characteristics of the geological formation and high spatial and temporal variability in rainfall, the reliability of the existing data on groundwater recharge is highly questionable. Therefore, attempts should be made to accurately estimate recharge from precipitation. Far more important is the requirement for robust methodologies for estimating the groundwater balance, which considers inflows and outflows from the aquifers. More importantly, the assessment of over-exploitation should consider the negative consequences of groundwater over-use, which are physical, economic, social, and ethical, rather than merely considering the abstraction against utilizable recharge.

As discussed in Chapter 3, this is one of the crucial problems hindering serious groundwater resource planning for agriculture in India. Current estimates seriously under play groundwater over-development problems in the hard rock regions of India (Kumar and Singh, 2008), as they fail to capture the undesirable effects, such as a sharp decline in water levels, rampant well failures and reduction in well yields, all of which have serious economic consequences (Kumar and Singh, 2008).

Groundwater resource estimation is a politically sensitive issue in India. Representatives of political parties want to put their constituency into the "safe" category, even when they are over-exploited or critically exploited. The reason is that most of the institutional financing for well development for an administrative unit is dependent on the status of groundwater development for that unit (Moench, 1992). As a result, the agency estimates are often subject to tampering, as people's representatives would like to see fund flows for drilling wells and the purchase of pump-sets in their constituency continue uninterrupted. It is important to remember that access to rural livelihoods is linked to access to groundwater. Therefore, groundwater resource estimation could be a big source of rent-seeking by official agencies. In order to make it free from political interference, it is important that the task of resource evaluation and planning is assigned to independent agencies.

The optimum development of groundwater would help minimize the cost of abstraction of unit volume of water. Therefore, mechanisms for regulating groundwater use will have to be thought through in some detail before groundwater use becomes popular and exploitation becomes intensive. If groundwater is also abundant in a region with a shallow water table, low cost irrigation systems such as treadle pumps could be promoted, taking advantage of the shallow water table conditions and rural labour force. Another option would be micro diesel pumps combined with very shallow tube wells. Nevertheless, the rich farmers might be able to install pump sets and use a larger volume of groundwater. In areas with extremely limited groundwater resources, the traditional wells with rope and bucket or treadle pumps would be appropriate if the water table is within 20–25 feet below the ground.

Planning surface irrigation systems for conjunctive management

Sub-Saharan Africa has a savanna climate, with erratic rainfall and moderate population pressures. The water use to water resource endowment ratio remains low due to lack of irrigation. Though, theoretically, there is a lot of unused potential, the water is difficult to mobilize due to problems of "coping capacity", which is due to poor institutional preparedness, lack of human resource capacity and financial constraints (Falkenmark and Rockström, 2004).

Within the tropical semi-arid region of sub-Saharan Africa, there are regional differences. Rainfall gradients are steep: as much as 100 mm per 100 km in West Africa. The rainfall ranges from 100 mm in the northern region of the Sahelo-Saharan zone to over 1,600 mm in the Guinean zone. The duration of the rainy season also varies greatly, ranging from one month in the desert margin to more than eight months in the Guinean coastal zone.

The Indian experience indicates that the focus should be on the balanced development of surface water and groundwater resources for irrigation in sub-Saharan Africa, failing which there could be several undesirable consequences. Some of them witnessed in India are water-logging and salinity caused by under-utilization of groundwater resources in the canal command areas (the result of the availability of cheap canal water against expensive groundwater exploitation); and the over-exploitation of groundwater in the semi-arid and arid regions, which results from a lack of surface water.

To facilitate this, action would be required at two levels, first at the policy level and second at the level of the planning, design and execution of irrigation schemes. As a matter of policy, naturally water-scarce arid and semi-arid regions should embark on surface irrigation using water imported either from water-rich catchments within the region or from water-abundant regions. Such imports might be possible in west Africa and, to an extent, in east Africa, due to the high rainfall gradients with movement of water from south to north. At the next level, the schemes, while providing irrigation water to crops in that region, would also simultaneously enable recharge of the local shallow aquifers through irrigation return flows. This would ensure sustainable groundwater use in such regions. By embarking on conjunctive management principles, the need for a high degree of water exploitation could be minimized, thereby reducing the chances of "technical water stress" which tropical countries in South Asia, including India, are facing in most areas.

As a matter of policy and practice, in the high rainfall regions of Kenya and Ethiopia and southern parts of western Africa, small water harvesting systems such as mini reservoirs should be promoted. Lift devices could be used to draw water from these reservoirs to provide supplementary irrigation to crops during the rainy season, at times of dryness. High rainfall over a longer duration also means a lower cost of harvesting unit volume of water. The region will not require full irrigation due to the availability of green water.

Planning water systems as multiple use systems

The cost of the exploitation of water resources is going to be very high in Africa for a variety of reasons (Foster et al., 2006; Rosegrant et al., 2001). Therefore, it is imperative that the values realized from their uses are higher than if the same is diverted for crop production alone.

At the same time, the willingness of people to pay for water services, be it for productive uses or domestic uses, is likely to remain low, due to the poor economic conditions and a lack of understanding of the benefits of using more water (in the case of productive uses like irrigation) or better quality water (in the case of domestic uses). In poor neighbourhoods, water systems which are not capable of meeting the multiple needs of the communities are unlikely to find importance in their day to day affairs, which would affect people's willingness to pay for the services being rendered (GSDA, IRAP and UNICEF, 2011).

A sufficient amount of literature now exists to show that a marginal increase in the volume of water supplied from a single use system or a marginal improvement in the quality of its water could result in significant gains in the economic benefits realized from its uses, and that these can far exceed the costs involved. Using data from Sri Lanka, Renwick (2008) showed that an increase in the water available from a rural water supply system to meet livestock rearing, kitchen garden and small enterprise needs resulted in an increased income of US $25 to US $70 per capita per year (US $1 equates to INR 50). If the amount of water supplied to rural households from a drinking water supply scheme is increased marginally beyond the domestic level, people might show greater willingness to invest in livestock rearing and kitchen gardens, thereby increasing their economic outputs and improving livelihoods. This would also increase their willingness to pay for drinking water supply services, and to maintain the systems.

Two distinct possibilities exist in sub-Saharan Africa. First, the physical infrastructure of small and large irrigation schemes could be extended to cover the rural water supply, given the fact that the access to improved water supply is quite low in this region. If the source of irrigation is surface water from rivers and lakes, the water might require some preliminary treatment using sand filters, etc., for removal of physical contaminants and organic matter, before being supplied for domestic uses. In the case of groundwater-based irrigation systems, they could be used for multiple purposes without much additional infrastructure.

On the other hand, if rural water supply schemes are built, then provision could be made to increase the total per capita supplies to accommodate productive uses such as livestock rearing, the kitchen garden and tree planting in the villages. In this case, the cost of production of unit volume of water would be less than it would be if water is only used for drinking and domestic purposes. In such cases, small reservoirs will have to be built at different points within the village for livestock drinking, if livestock demand is major. If kitchen gardens and homesteads are a priority for the community, additional connections will have to be provided so the households can take water directly into their domestic areas.

Farming system approach in promoting micro-irrigation

In view of the fact that the access to water for agriculture is very poor in sub-Saharan Africa, owing to the very high cost of exploitation, many scholars have advocated micro-irrigation systems as a technology to enable farmers produce to improve the food security situation and reduce poverty with only a limited amount of water (Postel et al., 2001; Keller, 2002). Low cost micro-irrigation systems, such as bucket kits and drum kits, were suggested for farmers with very small holdings and extremely limited access to water resources. But such suggestions lacked a farming system perspective, both at the level of individual farms and at the regional farming system level, which has to look at the risks associated with such farming models.

First of all, micro-irrigation systems are most amenable to fruit and vegetable crops and some of the cash crops, such as cotton, groundnut and potato. These technologies are not yet proven to be techno-economically viable for cereals, thus growing these crops will not help farmers to meet their staple food needs. They might be able to sell their produce in the market and purchase food grains from the earnings, but the earnings available from farming will depend on how successfully they are able to market these crops, and how backward and forward linkages are established. Large-scale production of these crops can lead to a price crash, if new markets are not developed in tandem, bringing tremendous hardship to the growers. Nevertheless, the issue is not limited to the marketing of produce alone. Most of these crops are highly susceptible to disease. Crop protection measures are extremely important for the survival of these crops and for good harvest. It is important to remember that the region still lags behind in terms of research-based knowledge on agriculture and extension services (Rosegrant et al., 2001).

Also, the input costs are high for fruits and vegetables, both in terms of seeds and pesticides. The farmers need to have sufficient capital. So far as fruits are concerned, there is an additional burden of the long gestation period: the minimum time duration for horticultural crops is one year for papaya, and can sometimes be up to three to four years, for crops such as sapota, guava, lemon and other citrus fruits. For farmers with marginal hand holdings, this might come as an added constraint. All these increase the farming risk.

An alternate scenario is that if large numbers of farmers from a region succeed in adopting new farming systems based on market-oriented crops with the use of micro-irrigation technologies, then this can even motivate them to replace the traditional cereal crops in their farms with the high valued cash crops for earning greater income. This was the trend found in the north Gujarat region, where large-scale adoption of micro-irrigation systems along with fruits and vegetables occurred, with some shrinkage in the area under cereals such as wheat and bajra. Similar trends can cause regional food shortages and food inflation in the context of sub-Saharan Africa, as the region is already heavily dependent on food imports.

For both the scenarios, institutional mechanisms are important in reducing the risk. In the first case, it will be in the form of proper extension services for agronomic inputs and credit services. In the second case, it will have to be in the form of market support, such as better support prices and procurement policies for cereals, in order to create an incentive for farmers to continue growing.

Development of market infrastructure for high valued crops

In sub-Saharan Africa, transportation and communications infrastructure is far more limited than it was when the Green Revolution began in Asia (Rosegrant et al., 2001). If the communities in large regions take up the cultivation of fruits (such as banana) and vegetables (bean) on a large scale, the dependence on export, as well as on local markets, would increase. Transportation and communication would be one of the greatest constraints for farmers in obtaining remunerative prices for their produce. First of all, farmers will have to transport the fast-perishing produce to the nearest market to fetch decent prices. Second, there will need to be proper communication facilities so that farmers can obtain adequate information about the price trends in various neighbouring markets in order to take timely decisions on harvest, storage, processing and marketing. But, as the FAO (2009) notes, given the very low population densities, any infrastructure connecting farmers to markets would be costly.

Post-harvest technologies for the farm produce would be yet another requirement if the export market is to be tapped. Adequate storage and processing facilities have to be created. Some African countries have already started tapping export markets to sell their produce; for example, Kenya exports flowers to Europe.

Priority areas for future research

Can groundwater be the future of Africa's agriculture?

One of the important areas for future research is the potential of groundwater for irrigation development and water supplies in developing countries. In India, there is a reasonable amount of geo-hydrological mapping already done, and groundwater resources are estimated; the challenge is in refining methodologies to realistically assess the degree of over-exploitation. Sub-Saharan African countries pose greater challenges in planning groundwater development, as the information available on aquifer recharge and abstraction are too little and patchy. The priority for countries in sub-Saharan Africa is to generate the knowledge and information about renewable groundwater and the stocks before they embark on policies for the large-scale development of this resource. The scientific challenge is great, in view of the greater climatic variability across space and across years, and the variation in geological formation.

Bio-fuel and food security linkage

Globally, there is increased demand for crops that can be used in bio-fuel (ethanol) production, with the result that, while the output of these is increasing, the diversion to bio-fuel production far exceeds the increase in production. For instance, in the

United States, while the production of maize increased from 200 million US tons to nearly 340 million tons between 1995 and 2010, more than 100 million tons now goes for bio-fuel production, whereas it was less than 10 million tons in 1995 (Earth Policy Institute, 2011). The bio-fuel mandate of several developed and emerging economies is exacerbating price fluctuations and increasing the volatility of global food prices, in addition to magnifying the tension between the demand and supply of food, threatening food security. Thus, efforts to achieve energy security might come at the cost of food security. In sub-Saharan Africa, the area cultivated is less than 25 per cent of the arable land (FAO, 2009). With rising international crude oil prices (IFPRI, 2011) and with many developed and developing countries having ambitious plans for bio-fuel production in the future, to replace fossil fuel, there is going to be enormous pressure on land, water and nutrients. There is growing evidence that, while the less productive land is likely to be targeted for this in the beginning, slowly the more productive land, which is the natural habitat for crops, could be allocated for bio-fuel production. The impact of this on food production, food supplies and cereal prices needs to be analyzed carefully.

How low cost are low cost irrigation devices?

The cost of the exploitation of water resources through conventional means is high in African countries. This, along with poor economic conditions, has hindered irrigation development in the continent. In eastern India, high poverty rates act as a constraint to improving water security through wells, though groundwater is abundant. One important determinant of irrigation expansion in sub-Saharan Africa and eastern India is the cost of irrigation. Therefore, low cost irrigation technologies have attracted immense attention during the past one-and-a-half decades or so. These include treadle pumps (Shah et al., 2000; Kumar, 2000) and low cost drip systems which work under low heads (Postel et al., 2001; Ngigi, 2008). There are limited studies on the socio-economic impacts of these technologies (Kumar, 2000; Shah et al., 2000, for treadle pumps).

But, in spite of the fact that, internationally, researchers have advocated this as the panacea for the land and water-starved poor farmers of Asia, Latin America and sub-Saharan Africa, with huge potential to transform agriculture and reduce food insecurity and hunger by raising crop yields and farmer incomes (Friedlander, undated; Postel et al., 2001), to date there are no scientific evaluation studies on their efficiencies, or on the actual potential of these technologies in different regions. For tradable pumps, energy use efficiency for human labour is extremely important if they have to cater to the needs of poor communities, who are also food insecure. Similarly, water use efficiency in low cost drip irrigation as against high-end precision irrigation (such as pressurized drips and sprinklers) needs to be ascertained through scientific studies. Experiments carried out in Kenya on the performance of various drips under low pressure heads show that the emission uniformity (EU) and flow variation (FV) were not within desirable limits even for small lateral lengths (of 15 metres), when land was sloping (more or less than 0.0 per cent slope) at 1.0metre pressure head. Again, the values of EU and FV were found to be unacceptable when the length of the lateral exceeded 15metres, even under 0.0 slope and 1.0 metre head (Ngigi, 2008). This basically means that such low head systems will work for tiny plots when fields are leveled. Flow variation and emitter uniformity would have a significant effect on the crop growth and, hence, water use efficiency.

Conclusions

There is no magical wand or silver bullet for turning around agriculture in developing economies. The challenges are far greater in the poor income countries of sub-Saharan Africa. That said, wise management of water is going to be crucial for food security and sustainable agriculture. The billion dollar question is what water management for agriculture really means in the context of these developing economies, which are mostly falling in semi-arid, arid and, sometimes, humid tropics. There is a need to understand climatic variability and its implications for the paradigm of water resource development and water management.

Large water resource systems, which are capable of transferring water from abundant regions to scarce regions, would play a crucial role in improving water security, boosting agricultural production and reducing poverty in the rural areas of poor developing economies, particularly those in the semi-arid to arid tropics which also experience high variability in rainfall and low dependability of streamflows. In such regions, groundwater alone will not be able to meet all the irrigation water demands in the long run. In contrast, the small water harvesting systems, along with the utilization of shallow groundwater, would be more viable in high rainfall, humid, tropics, which experience low variability in rainfall.

On the other hand, there is a need for greater recognition of the opportunities provided and constraints induced by the "arable land-water resources balance" in different regions, as they would determine the success and effectiveness of different water resource development paradigms and strategies. Scarcity of arable land would be a significant constraint to boosting food production in certain regions, even through water resources are plenty. In contrast to this, in certain other regions, vast tracts of arable land are left fallow due to an acute shortage of water. This makes the transfer of water from surplus regions to scarce regions inevitable to increasing irrigation potential. The fact that most of the food exporting regions in the world are naturally water-scarce regions, and those which import food are naturally water-rich regions (Kumar and Singh, 2005), should force us to pay attention to this nexus between land, water and food.

The water management alternatives for regional agriculture should be determined by hydrology, geo-hydrology, climate and soils, along with the socio-economic dynamic, such as access to land and capital, cropping systems, traditional knowledge about farming, the labour supply, and the institutional set-up (such as agricultural extension services and access to credit and market support), and, therefore, the choices would be limited. Yet, judicious water management is what could change many of these input parameters, by removing the constraints and expanding choices to enhance the overall developmental outcome. Therefore, water management interventions for economically poor regions of Africa, which have been practising subsistence agriculture for generations, have to be slow, so as to allow us to understand how the farming community adapts to the changes.

Notes

1. Such practices include zero tillage technology, plastic mulching and alternate wetting and drying for paddy.
2. Here, water security does not refer to water for agricultural production alone, but it refers to water for basic survival needs, water for livestock and water for crop production.

References

Barrios, S., Bertinelli, L., & Strobl, E. (2004). Rainfall and Africa's growth tragedy. Unpublished paper.

Batchelor, C., Singh, A., Rao, R.M., & Butterworth, J. (2002). Mitigating the potential unintended impacts of water harvesting. Paper delivered at the IWRA international regional symposium on water for human survival, New Delhi, India, 26–29 November.

Bhattarai, M., & Narayanamoorthy, A. (2003). Impact of irrigation on rural poverty in India: An aggregate panel data analysis. *Water Policy, 5*(2003), 443–458.

Biswas, A. K., & Tortajada, C. (2001). Development and large dams: A global perspective. *International Journal of Water Resources Development, 17*(1).

Dhawan, B. D. (2000). Drip irrigation: Evaluating returns. *Economic and Political Weekly, 35*(42), 3775–3780

Earth Policy Institute. (2011). Data center: Climate, energy, and transportation. http://www.earth-policy.org/data_center/C23.

Falkenmark, M., & Rockström, J. (2004). *Balancing water for humans and nature: The new approach in eco-hydrology.* New York, USA: Earthscan.

Food and Agriculture Organization (FAO). (2006). *The AQUASTAT database.* Rome, Italy: Food and Agriculture Organization. www.fao.org/aq/agl/aglw/aquastat/dbase/index.stm.

Food and Agriculture Organization. (2007). *Coping with water scarcity: Challenges of the 21st century.* A World Water Day 2007 Brochure. Rome, Italy: Food and Agriculture Organization.

Food and Agriculture Organization (FAO). (2009). *Report of the FAO expert meeting on how to feed the world in 2050.* Rome, Italy: Food and Agriculture Organization.

Foster, S., Tuinhof, A., & Garduño, H. (2006). *Groundwater development in sub-Saharan Africa: A strategic overview of key issues and major needs.* Case profile collection 15. Washington, DC, USA: The World Bank.

Friedlander, L. (undated). *Drip irrigation in sub-Saharan Africa: Toward a strategy for technology transfer. Overview and preliminary recommendations.* Sede Boqer, Israel: The Jacob Blaustein Institutes for Desert Research, Ben Gurion University of the Negev.

Gommes, R., & Petrassi, F. (1994). *Rainfall variability and drought in sub-Saharan Africa since 1960.* Agro-meteorology working paper no. 9. Rome, Italy: Food and Agriculture Organization of the United Nations.

Groundwater Survey and Development Agency, Institute for Resource Analysis and Policy and UNICEF. (2011). *Multiple use water services to reduce poverty and vulnerability to climate variability and change: Feasibility report of a collaborative action research project in Maharashtra, India.* Mumbai, India: UNICEF.

Hussain, I., & Hanjra, M. (2003). Does irrigation water matter for rural poverty alleviation? Evidence from South and South East Asia. *Water Policy*, *5*(5), 429–442.

International Food Policy Research Institute. (2011). *2011 Global Hunger Index: The challenge of hunger-taming price spikes and excessive food price volatility*. Washington, DC, USA: International Food Policy Research Institute, Concern Worldwide and Welthungerhilfe.

Kumar, M. D. (2000). *Irrigation with a manual pump: Impact of treadle pump on farming enterprise and food security in coastal Orissa*. Working paper 148. Anand, Gujarat, India: Institute of Rural Management Anand.

Kumar, M. D. (2005). Impact of electricity prices and volumetric water allocation on groundwater demand management: Analysis from western India. *Energy Policy*, *33*(1), 39–51.

Kumar, M. D. (2007). *Groundwater management in India: Physical, institutional and policy alternatives*. New Delhi, India: Sage Publications.

Kumar, M. D. (2009). *Water management in India: What works and what doesn't*. New Delhi, India: Gyan Books.

Kumar, M. D., & Singh, O. P. (2001). Market instruments for demand management in the face of scarcity and overuse of water in Gujarat. *Water Policy*, *5*(3), 387–403.

Kumar, M. D., & Singh, O. P. (2005). Virtual water in global food and water policy making: Is there a need for rethinking? *Water Resources Management*, *19*, 759–789.

Kumar, M. D., & Singh, O. P. (2008). How far is groundwater over-exploited in India? A fresh investigation into an old issue. In M. D. Kumar (Ed.), *Managing water in the face of growing scarcity, inequity and declining returns: Exploring fresh approaches*. Proceedings of the 7th annual partners' meet, IWMI-Tata Water Policy Research Program, ICRISAT, Patancheru, 2–4 April, 2008.

Kumar, M. D., & van Dam, J. (2009). Improving water productivity in agriculture in developing economies: In search of new avenues. In M. D. Kumar, & U. Amarasinghe (Eds.), *Water productivity improvements in Indian agriculture: Potentials, constraints and prospects*. Strategic analysis of national river linking project of India, series 4. Colombo, Sri Lanka: International Water Management Institute.

Kumar, M. D., Ghosh, S., Patel, A., Singh, O. P., & Ravindranath, R. (2006). Rainwater harvesting in India: Some critical issues for basin planning and research. *Land Use and Water Resources Research*, *6*(2006), 1–17.

Kumar, M. D., Patel, A., Ravindranath, R., & Singh, O. P. (2008). Chasing a mirage: water harvesting and artificial recharge in naturally water-scarce regions. *Economic and Political Weekly*, *43*(35), 61–71.

Moench, M. (1992). Drawing down the buffer. *Economic and Political Weekly*, *27*(13), A7–A14

Ngigi, S. N. (2008). Technical evaluation and development of low head drip irrigation systems in Kenya. *Irrigation and Drainage*, *57*, 450–462.

Ong'ayo, O. A. (2008). Political instability in Africa: Where the problem lies and alternative perspectives. Paper presented at Symposium 2008 – Afrika: een continent op drift. Stichting Nationaal Erfgoed Hotel De Wereld, Wageningen, 19 September.

Oweis, T. A. H., & Kijne, J. (1999) Water harvesting and supplementary irrigation for improved water use efficiency in dry areas. Colombo, Sri Lanka: International Water Management Institute.

Postel, S., Polak, P., Gonzales, F., & Keller, J. (2001). Drip irrigation for small farmers: A new initiative to alleviate hunger and poverty. *Water International*, *26*(1), 3–13.

Ray, S., & Bijarnia, M. (2006). Upstream vs downstream: Groundwater management and rainwater harvesting. *Economic and Political Weekly*, *41*(23), 2375–2383.

Renwick, M. (2008). Multiple use water services. Presentation given to the GRUBS Planning Workshop, Nairobi, Kenya, November.

Rockström, J., Barron, J. and Fox, P. (2002) Rainwater management for improving productivity among small holder farmers in drought prone environments, *Physics and Chemistry of the Earth*, 27, pp 949–959.

Rosegrant, M. W., Paisner, M., Meijer, S., & Witcover, J. (2001). *2020 global food outlook: Trends, alternatives and choices. A 2020 vision for food, agriculture, and environment initiative*. Washington, DC, USA: International Food Policy Research Institute.

Scott, C., & Ochoa, S. (2001). Collective action for water harvesting irrigation in Lerma-Chapala basin, Mexico. *Water Policy, 3*, 55–572.

Shah, T. (2009). *Taming the anarchy: Groundwater governance in South Asia*. Colombo, Sri Lanka: Resources for Future and the International Water Management Institute.

Shah, T., Alam, M., Kumar, M. D., Nagar, R. K., & Singh, M. (2000). *Pedaling out of poverty: Social impact of a manual irrigation technology in South Asia*. Research Report 45. Colombo, Sri Lanka: International Water Management Institute.

Shah, Z., & Kumar, M. D. (2008). In the midst of the large dam controversy: Objectives and criteria for assessing large water storages in the developing world. *Water Resources, 22*, 1799–1824.

Siebert, S., Burke, J., Faures, J. M., Frenken, K., Hoogeveen, J., Döll, P., & Portmann, F. T.(2010). Groundwater use for irrigation: A global inventory. *Hydrology and Earth System Sciences, 14*, 1863–1880.

Syme, G., Reddy, V. R., Pavelik, P., Croke, B., & Ranjan, R. (2011). Confronting scale in watershed development in India. *Hydrogeology Journal*. http://dx.doi.org/10.1007/s10040-011-0824-0.

United Nations. (2006). *Human development report 2006*. New York, USA: United Nations.

United Nations. (2009). *Human development report 2009. Overcoming Barriers: Human mobility and development*. New York, USA: United Nations.

United Nations Development Programme. (2006). *United Nations Development Programme report*. New York, USA: United Nations Publishing.

von Braun, J. (2007). *The world food situation: New driving forces and required actions*. Washington, DC, USA: International Food Policy Research Institute.

Vyas, J. (2001). Water and energy for development in Gujarat with special focus on the Sardar Sarovar project. *International Journal Water Resources Development, 17*(1), 37–54.

Wiesmann, D. (2006). *Global hunger index: A basis for cross country comparison*. Washington, DC, USA: International Food Policy Research Institute.

You, L., Wood, S., & Sichra, U. W. (2007). *Generating plausible crop distribution and performance maps for sub-Saharan Africa using a spatially disaggregated data fusion and optimization approach*. Discussion paper 00725. Washington, DC, USA: Energy and Production Technology Division, International Food Policy Research Institute.

Index

Accelerated Irrigation Benefit Programme 164
Africa: food security 1, 4; irrigation development 219; micro-irrigation 135; water shortage 4
agriculture: capital formation 195; development policies 184; exports 188; GDP from 189; growth strategy 164; holdings fragmentation 197; mechanization 196; policy 195; production 38, 164–80; productivity 13, 41, 195; public investment 185, 194; subsidies 195–6; water use 174
agronomic practices 169
Andhra Pradesh: bore wells 18–19; green revolution technologies 42; ground water banking 173; ground water exploitation 7, 26, 48, 50–2; irrigation management transfers 77; net irrigated area 172; tank irrigation 30; water shortage 171; water users' associations 84
aquifer: yield 44–5, 61; hard rock 18, 44, 173; phreatic 61
arable land 54; land-water balancing 225; per capita 18, 32
arid region: water recharge 17
artificial recharge structures 27
Asia: agricultural exports 188; food production needs 199; food security 1; water scarcity 4; agricultural development 184–204
Assam: irrigation development 167

Bangladesh: agricultural expenditure 194; agricultural methane emissions 200; economic growth 189; irrigation reforms 76; multiple use water systems 161; water management 22; water resource development 186; water users' association 93
Bihar: irrigation development 2, 32, 41, 167; water allocation 175
bio-fuel crops 14, 224
bore wells 46; Andhra Pradesh 18; yield 18
Brahmaputra basin 40

canal irrigation 5, 76–93; engineering efficiency 25; pricing 56–7, 176; water diversion 29
Cauvery river basin 24
cereal production 6, 38; per capita 166
Chattisgarh: irrigation management transfer 77
China: agriculture development 194; large dams 61; irrigation reforms 76; water users' association 93
climate change 200
Columbia: irrigation reforms 76
commercial crops 187–8
common property rights 139, 174
cropping intensity 42; climate change 200
CROWPAT model 18

dairy farming 38
dams: displacement 66; indirect benefits 212; large vs small 7; reservoir 65; storage 64
Dharoi irrigation command 29, 82
Diggie (water storage) 210
downstream social conflict 26
drinking water 3, 29; rural 12; sub-Saharan Africa 214; urban 30
drip irrigation 169; benefit–cost ratio 129

eco-system conservation 185

energy crops *see* bio-fuel crops
environmental degradation 200
Ethiopia 161
evapo-transpiration 18, 24, 32,44; Rajasthan 98

farm mechanization 196
farming system 116–35
fertilizer subsidies 196
flood control 165
food exporting regions: water availability 225
food grains consumption pattern 3; per capita availability 38, 166
food production 41, 51, 187; Asia 199; per capita 191–2
food security 1–14, 5, 7, 38–57, 166, 184, 191, 198; developing countries 208–26; sub-Saharan Africa 213
fresh water withdrawal 193; Ganga basin 22; Brahmaputra basin 22; Meghna basin 22; geo-hydrological mapping 223

Global Hunger Index 7, 214–15
Godavari delta 30
gravity irrigation 171
green revolution 1, 13, 185–6, 195
green water utilization 55
gross irrigated area 20, 49; wells 28
groundwater: abstraction 21, 47; balance 55; banking 26, 173; benefit–cost 21, 47; depletion 7, 17, 20; GDP contribution 49; irrigation potential 7, 54; Rajasthan 101; social impact 172; stock 45, 211, 219
groundwater development 3; Africa 219; private sector 216; Rajasthan 100
groundwater exploitation 10, 39, 40–1, 43, 50, 54–6; Andhra Pradesh 48; ethical aspects 48; food production 51; irrigation potential 54; Madhya Pradesh 48; negative aspects 56; north Gujarat initiative 10; per capita 18, 20, 50; regulation 219; rights 29; state-wise comparison 33; strategy in Gujarat 117–18; Tamil Nadu 48
groundwater pumping: economic cost 171
groundwater recharge 8, 12, 26–7, 43, 53; Gujarat 28–30
Gujarat: borewells 18–19; canal irrigation 5; green revolution technologies 42; ground water exploitation 48, 50–2, 169, 171, 196; ground water recharge 28–30; irrigation management transfer Performance 77, 79, 81; irrigation sector reforms 9–10; power tariff 177–8; renewable water 118; seasonal irrigation water tariff 81; surface irrigation systems 25–6; water management 116–35

Haryana: Green Revolution 1–2; ground water exploitation 7, 18–19; green revolution technologies 42; ground water exploitation 51, 169, 196; power tariff 26, 177
high yielding varieties 42, 186
Himachal Pradesh 25
horticultural crops 56, 169, 170; micro-irrigation 222
hydroscopic water 119

Indira Sagar dam 69
Indo-Gangetic plains 2
inter-basin water transfer 57, 172
inter-sectoral water demand 139
Iran: irrigation management transfer 93
irrigated area: Andhra Pradesh 172; developing countries 76
irrigated crop production returns 27
irrigation 17–34; benefit–cost analysis 41; cost of 224; cost per hectare 165; development: misconceptions 17; devices 224; donor funding 78; economic value 145; farmers management 84; infrastructure 195; public–private partnership 93, 201; investment 4, 185–6, 214; maintenance 9; participatory management 199; physical productivity 218; policies 209; potential 78; power consumption 21; quality and reliability 176; reforms 76; revenues 9, 85, 89; and rural poverty 214; service in Gujarat 51; social benefit 26; subsidies 196, 212; technology 119, 167; in Tenth Five-Year Plan 165; water charges 81; in Maharashtra 89; water rationing 176; water use for crops 123
irrigation command: Dharoi 29, 82; Mahi 25, 29–30; Mulla 25, 29; Ukai-Kakrapar 25
irrigation management transfer 76–7, 91, 199; legislation 77; NGOs 88; performance of 79

Jharkhand 167

Karnataka: ground water banking 173; ground water exploitation 48, 50, 196; water shortage 171–2; well irrigation 26
Kenya: multiple use water systems 161
khadins 9,102; benefit–cost analysis 10, 105; crop economics 104; hydrological impact 107; socio-economic impact 103, 109–11
Krishna river basin 24–5

land fragmentation 13, 42,190
large dams 61–73; benefit–cost analysis 8, 22; definition 62, 64, 65; displacement 65–8; economic benefit 69–70; environmental impact 61; evaluation 61–73, 212; growth 63; net social welfare 72; performance 67; salinity 68; socio-economic impact 65
Lerma Chapala basin 218
Luni river basin 100

Madhya Pradesh: canal irrigation 5, 9; ground water exploitation 48, 196; irrigation management transfer 77, 83–5, 91; irrigation Reforms 9; water shortage 171; well irrigation 22, 26, 47
Mahanadi basin 141
Maharashtra: canal irrigation 5; hard rock acquifers 46; irrigation management transfer 77, 87–91; irrigation reforms 9; surface water irrigation 25–6; water shortage 171
Mahi irrigation command 25, 29–30
Mali 76
micro diesel pump 167, 179, 219
micro irrigation 4, 5, 55, 169–70, 209–11; Africa 135; applied water saving 132,133; benefit–cost analysis 130, 131, 133; devices 118; farming systems approach 222; Gujarat 10, 116; horticultural crops 222; real water saving 132; tanks 138; water productivity 133
minor irrigation 165; crop yield 125; impact on livestock 127
monsoon 43, 173; flood water 173
Mulla irrigation command 25, 29

Narmada river basin 22, 24, 27, 43, 47
Nepal: multiple use water systems 161

North Gujarat Groundwater Initiative 10, 119
NREGA 12, 164, 170; negative impact 166

open wells 171
Odisha: irrigation development 2, 167–8; irrigation management transfer 77; multiple use water systems 5, 12, 138–62; water allocation 175–7

Pakistan: irrigation management transfer 76, 93
Pennar river basin 24
Philippines 76
post-harvest technologies 223
power tariff 6, 42, 53, 57, 132, 177–9, 196, 209, 212; slab system 179
power supply: quality of 177
public–private partnership: irrigation infrastructure 201
pump irrigation subsidies 26, 53
Punjab: arable land 32; borewells 18–19; Green Revolution 1–2; green revolution technologies 42; ground water exploitation 7, 48, 50–1, 169, 196; power tariff 177; pump irrigation 26

rain-fed crops 5, 12
Rajasthan: groundwater depletion 19, 46, 50–1, 101, 196; groundwater recharge 26; surface water irrigation 25–6; water harvesting, traditional method 97–114; water shortage 171
regional water balance 39
renewable water: Gujarat 118; Ganga-Brahmaputra basin 23
reservoir: sedimentation 30; capacity 63
rice: drought prone variety 200
river basins: Cauvery 24; Ganga 22–3; Brahmaputra 22–3; Meghna 22; Krishna 24–5; Lerma Chapala 218; Luni 100; Mahanadi 141; Narmada 22, 24, 27, 43, 47; Pennar 24; Sabarmati 100
river lifting 31
rural innovations 197
rural livelihood, Asia 190

Sabarmati river basin 24
Sardar Sarovar dam: 28–9, 61, 69; indirect benefits 212
siltation 18, 30

small and marginal farmers 5, 21
soil water depletion 55
South Asia 7, 76
South Korea: irrigation development 186–9; arable land 197
sprinkler irrigation 169; benefit–cost analysis 129
Sri Lanka: irrigation management transfer 93; multiple use water systems 161, 221; agricultural development 194
Sub-Saharan Africa: food security 213–17; surface irrigation systems 220–1; water harvesting systems 113; water resource development 208–9; water scarcity 4, 12–14
Sudan irrigation reforms 76
surface irrigation 3–6, 17, 55, 220; benefit–cost analysis 28
surface water: appropriation of 210; export 7, 29, 44, 213
sustainable agriculture 1–16; Asia 184; developing countries 208; strategies 198; technology options 184–204
sustainable development strategies 202
sustainable ground water use 167
sustainable water use index 215

Tamil Nadu: ground water exploitation 7, 19, 48, 50, 52, 196; green revolution technologies 42; ground water banking 173; irrigation management transfer 77; power tariff 26
tank eco-system 142; domestic water supply 153–4
tank irrigation: economic value of water 160; fisheries production 153–4; multiple use water systems 138–62; net return from crops 149–52; Odisha 138–40, 157; total economic value 145; water productivity 158
tank management 139
technical water stress 217, 220
techno economic water scarcity 218
Thailand: farm mechanization 197; food production 186–7, irrigation development 194; multiple user water system 161
total factor productivity 1, 2
treadle pump 167–8,179
tube well 18; average command area 47; Gujarat 52; hard rock areas 52; lifetime 155; poor discharge 47
Tunisia: Irrigation Reforms 76

Ukai-Kakrapar command 25
urban population growth 3
urbanization 173, 185

Uttar Pradesh: Green Revolution 1–2; tube wells 19; water allocation 175; water availability 41, 169

Vietnam 186–7, 197
virtual water trade 54
volumetric water allocation 57

Waghad dam 90
water: abstraction 46, 54, 138; cost per cubic meter 166; economic value 171; environmental concerns 4; governance regime 201; livestock drinking 11 ; logging 8,220; optimum use of 144; sustainable use 179; urban areas 4
water allocation regulatory approach 174
water delivery: infrastructure for 165
water demand management 118–19; Ganga-Brahmaputra basin 23
water efficient crops 4, 10
water harvesting 17,165; benefit–cost analysis 218; capital cost 218; community initiatives 210; incremental returns 218; per capita 18; sub-Saharan Africa 113; traditional methods 97–114
water intensive crops 3, 30, 173
water management 116–35; agriculture 5, 208–24; benefit–cost analysis 221; demand–supply balance 4; developing countries 208; farming system approach 116; Gujarat 116; strategies 202, 217; volumetric measure 175
water policy debate 32
water poverty index 14
water production cost 46
water productivity 55, 160; agriculture 167; net 118, 145; physical 129; technologies 168
water pumping energy cost 169
water regulation strategies 177
water replenishment strategies 180
water rights trading 174–5, 180
water saving technology 124 , 179
water scarcity 3–4, 40
water sector governance 193
water security 1, 4, 192; benefits 214
water services pricing 221

water sharing conflicts 161
water supply planning: economic benefits 221
water tariff 57, 175–6, 209, 221; Gujarat 81; recovery in Madhya Pradesh 87; willingness to pay 11
water use incentives 174
water use efficiency 3, 28, 54, 224
water use policy 164
water users' associations 9, 76–7, 82,176; alternative models 92; Andhra Pradesh 84; financial performance 91; Gujarat 80; international experience 92; Madhya Pradesh 84, 86; Maharashtra 87, 89; performance 84
well construction cost 52
well discharge 47
well failures 45–6
well irrigated area 4, 6, 20–1, 31, 49, 216; impact on poverty 31; average command area 21, 22, 47; drilling cost 48; optimum use 21; Madhya Pradesh 22
West Bengal: water resource development 2, 167; water allocation 175–7; power tariff 179
"wine and melon" systems 211
World Bank: assistance 194; irrigation development 76